The Development of the Laboratory

For Joasia

ISBN 0–88318–613–6

Printed in Hong Kong

This edition first published in the
United States and Canada in 1989 by the
AMERICAN INSTITUTE OF PHYSICS
335 East 45th Street, New York, NY 10017–3483

ISBN 0–88318–613–6

Printed in Hong Kong

Contents

List of Illustrations

List of Tables

Notes on Contributors

Lawrence Aronovitch is a doctoral candidate in Political Science at the Massachusetts Institute of Technology, where his work lies primarily in the field of science and technology policy. His dissertation will focus on the Japanese space programme. He is also a tutor in the History of Science at Harvard University.

Address: Center for International Studies, Massachusetts Institute of Technology, Building E38, Room 600, Cambridge, Massachusetts 02139, USA.

W. H. Brock is Reader in History of Science and Director of the Victorian Studies Centre at the University of Leicester. He has written extensively on the history of chemistry, the development of scientific education and the history of scientific periodicals.

Address: Victorian Studies Centre, University of Leicester, Leicester LE1 7RH, England.

David Cahan is Associate Professor of History at the University of Nebraska. He has recently published *An Institute for an Empire: The Physikalisch-Technische Reichsanstalt, 1871–1918*. His academic interests are the history of physics and the social history of science from the Enlightenment to the present.

Address: Department of History, University of Nebraska-Lincoln, 612 Oldfather Hall, Lincoln, Nebraska 68588-0327, USA.

Isobel Falconer is a Visiting Research Fellow of the Royal Institution Centre for the History of Science and Technology. She has made a special study of

J. J. Thomson and his attitude to experiment, and is currently investigating his influence on his research students. She was previously curator of the museum at the Cavendish Laboratory.

Address: Royal Institution Centre for the History of Science and Technology, The Royal Institution, 21 Albemarle Street, London W1X 4BS, England.

David Fenby is a physical chemist with research interests in the thermodynamics of liquid mixtures and isotope effects and the history of science. His historical work is concerned with the history of thermodynamics and with science in the nineteenth century.

Address: Department of Chemistry, University of Otago, PO Box 56, Dunedin, New Zealand.

J. Z. Fullmer is Professor Emeritus of History at Ohio State University. She is currently working on a biographical study of Humphry Davy.

Address: Department of History, Ohio State University, Columbus, Ohio 43210, USA.

Brian Gee was formerly Senior Lecturer in Physics at the College of St. Mark and St. John, Plymouth. He is currently working on aspects of the history of scientific instruments in the nineteenth century.

Address: 7 Barton Close, Landrake, Saltash, Cornwall PL12 5BA, England.

David Gooding is Senior Lecturer in the Science Studies Centre at the University of Bath and is a Visiting Research Fellow of the Royal Institution Centre for the History of Science and Technology. He is co-editor (with Frank A. J. L. James) of *Faraday Rediscovered* and (with Trevor Pinch and Simon Schaffer) of *The Uses of Experiment.* His study of the experimental rendering of natural phenomena, *The Making of Meaning*, is to be published shortly. He is writing a biography of Faraday, *Nature's Apprentice.*

Address: Science Studies Centre, School of Social Sciences, University of Bath, Claverton Down, Bath BA2 7AY, England.

Frank A. J. L. James is Lecturer in History of Science at the Royal Institution. He has written on nineteenth-century physics, chemistry, astronomy and belief, including editing (with David Gooding) *Faraday Rediscovered* and also *Chemistry and Theology in Mid-Victorian London: The*

Diary of Herbert McLeod, 1860–1870. He is currently editing the complete correspondence of Michael Faraday (1791–1867).

Address: Royal Institution Centre for the History of Science and Technology, The Royal Institution, 21 Albemarle Street, London W1X 4BS, England.

John Krige moved to Geneva in 1982 to work in an international group writing the history of CERN. Two volumes of this history have been published. He is at present leading a project which aims to cover the life of CERN up to the late 1970s. Previously he taught history and philosophy of science at the University of Sussex.

Address: Study Team for CERN History, c/o CERN, CH-1211 Geneva, Switzerland.

Dominique Pestre is *chargé de recherche* in the French CNRS. He originally worked on French science, on which he published *Physique et Physiciens en France, 1918–1940.* He is now working on the History of CERN.

Address: CNRS–CERN, 107 Rue du Temple, 75003 Paris, France.

Andrew Pickering is an Associate Professor in the Department of Sociology and Program in Science, Technology and Society at the University of Illinois at Urbana-Champaign. The author of *Constructing Quarks: A Sociological History of Particle Physics*, he was a member of the Institute for Advanced Study in Princeton in 1986–7, and an Exxon Fellow in the Massachusetts Institute of Technology Program in Science, Technology and Society in 1984–5. Before that he was at the Science Studies Unit of the University of Edinburgh. He is currently developing a pragmatist model of scientific practice.

Address: Department of Sociology, University of Illinois, 326 Lincoln Hall, 702 S Wright Street, Urbana, Illinois 61801, USA.

Catherine Westfall received her PhD in the History of Science from Michigan State University in 1988. In addition to teaching history of science and technology courses at Michigan State University, she is currently working on two books — a technical history of wartime Los Alamos and a history of the founding and first two decades of research at Fermilab.

Address: Lyman Briggs School, Michigan State University, East Lansing, Michigan 48823, USA.

Mari E. W. Williams is a lecturer in the Department of Economic History at the London School of Economics and a Visiting Research Fellow of the Royal

Institution Centre for the History of Science and Technology. After writing a thesis at Imperial College, London, on the history of astronomy, she worked on the history of the oil industry, and is currently completing a book on the history of the precision instruments industry in Britain and France since the late nineteenth century.

Address: Royal Institution Centre for the History of Science and Technology, The Royal Institution, 21 Albemarle Street, London W1X 4BS, England.

Foreword

In 1808 Humphry Davy laid before the Managers of the Royal Institution a paper that contained the following paragraphs:

> A new path of Discovery having been opened in the agencies of the Electrical Battery of Volta, which promises to lead to the greatest improvements in Chemistry and Natural Philosophy, and the useful Arts connected with them, and the increase of the size of the Apparatus being necessary for pursuing it to its full extent, it is proposed to raise a Fund by subscription, for constructing a powerful Battery, worthy of a National Establishment, and capable of promoting the great objects of Science.
>
> Already in other Countries, public and ample means have been provided for pursuing these investigations. They have had their origin in this Country, and it would be dishonourable to a nation so great, so powerful, and so rich, if, from the want of pecuniary resources, they should be completed abroad.
>
> An appeal to enlightened individuals on this Subject can scarcely be made in vain. It is proposed that the Instrument and apparatus be erected in the Laboratory of the Royal Institution, where it shall be employed in the advancement of this new department of Science.

Up to that time experimental scientists, by and large, had fashioned their own equipment and had worked in laboratories that were created from their own resources. Davy's appeal to the public for provision of laboratory equipment marked a new departure. His example is nowadays widely followed throughout the world. But even a century later there were many scientists who chose to establish their laboratories in their own homes using their own financial resources. A famous example was Lord Rayleigh, who, after establishing one of the foremost laboratories of physics in the world — the Cavendish Laboratory of the University of Cambridge — opted to set up a laboratory in his baronial home at Terling, Essex. The British Prime Minister (from 1885 to 1892 and again from 1895 to 1902) — the third Marquess of Salisbury — who was related by marriage to Lord Rayleigh, also set up a laboratory in his home at Hatfield, where he worked on magnetism and spectroscopy.

Other examples abound, as we learn from this fascinating book. Indeed, well into the twentieth century a few notable individual scientists, typified by the Cambridge applied mathematician G. I. Taylor (1886–1975), ran their home laboratories in parallel with larger ones in public places of learning and research.

The whole ethos of a laboratory is central to the understanding of the essence of modern science. This book will serve as a valuable guide to the central role that laboratories play in our industrial civilization.

The Royal Institution of Great Britain   J. M. Thomas, FRS
London, October 1988   Director and
                                    Fullerian Professor of Chemistry

Preface

These essays have been chosen to illustrate the development of the laboratory from its chemical beginnings to its present-day grandeur represented by the high-energy physics laboratories. They were also selected to illustrate the wide variety of influences that meet and interact in the architectural space called the laboratory.

Most, but not all, of these papers were presented to a Royal Institution Centre for the History of Science and Technology three-day symposium held from 17 to 19 September 1986 under the title *Laboratories: The Place of Experiment*. I thank all the contributors for substantially rewriting their essays following the ideas that emerged at the symposium and also with the subsequent selection criteria in mind. This has helped to bring some coherence to a subject whose chief characteristic is diversity.

Royal Institution, London, 1988   F. A. J. L. J.

Introduction

Frank A. J. L. James

A building set apart for conducting practical investigations in the natural sciences, originally and especially in chemistry for the elaboration or manufacture of chemical, medicinal, and like products (*OED*)

That laboratories are now fundamental to the practice of science is a commonplace. Yet the paucity of serious historical analysis of the subject, when compared with other areas of study, suggests that historians have not paid as much attention to this 'building set apart' as it deserves.[1] There have, of course, been 'in-house' accounts of individual laboratories, but very few of these have addressed more general problems and even fewer have employed recent historiographical tools in understanding the nature of laboratories.

There are perhaps two main reasons for this neglect of the laboratory as a historical phenomenon worthy of investigation. These are related to the complex reality of laboratories as they exist, which belies the apparent simplicity of laboratories expressed by the dictionary definition of the word. The dictionary definition implies that a laboratory is simply a building; but this is misleading. The word has at least three different meanings: it can be a building set apart, or a room within a building — the whole of which is not necessarily designated as a laboratory. It may also have a metaphorical meaning, especially when applied to the non-physical sciences, and from an early period after the first English usage of the word in 1605 it was indeed used in such a manner. The dictionary definition has, of course, changed over time. Most historians take the simple definition of 'laboratory' at face value and interpret what they find through such definitions. Such a viewpoint makes laboratories intrinsically uninteresting and unworthy of study. But as these essays illustrate, the word 'laboratory', which at first sight appears so transparent in its meaning, in fact represents a richly complex entity.

Laboratories did not exist in the pre-industrial age. Thus, it is no coincidence that the first English usage of the word appears in the early seventeenth century. We shall see that in many cases, patronage of laboratories is strongly linked to industrialization and this becomes increasingly so in more recent times. It is true that workshops existed before the seventeenth century and that laboratories can be viewed as a specialist development of them. But as a place where the exploration of natural phenomena was the sole task, together with all that that drew in its wake, laboratories are very much a product of, and a symbol of, modern industrial society. Its symbolic power is demonstrated by the wide variety of metaphorical usages that the word enjoys. That it is a poorly studied symbol is equally evident. The essays that are collected here show the range and diversity both of laboratories themselves and of the ways in which they can be studied. They are presented to stimulate further study of this place which is symbolically and materially so close to the centre of industrial culture and yet so very far from being 'set apart' from that culture.

In real life, away from dictionary definitions, there exist a wide variety of laboratories serving a multiplicity of interests and purposes. Laboratories are used for teaching, for research, for establishing standards, for technological and commercial purposes. Laboratories develop through a variety of stages. Funding has to be obtained to build and subsequently maintain them; they have to be designed; staff (including researchers, technicians and maintenance personnel) have to be recruited and supported. Experiments and observations have to be made and interpreted. The location and structure of the building as an architectural space may produce profound effects in all these aspects. Scientific results have to be published and the general perception of the trustworthiness of these is usually affected by the prestige, or otherwise, of the laboratory where the work was performed. Hence, it becomes clear that rather than a laboratory being a 'building set apart', it becomes a point at which a variety of social networks meet and interact to produce knowledge and trained scientists. This is a dynamic process which changes the character of a laboratory over time. Such processes are difficult for any single historian to untangle and this, too, has perhaps discouraged serious analysis.

The wide variety of laboratories that exist, and the multiplicity of their development, invites and allows many different approaches to the subject. This diversity is reflected in the contributions to this book. While it would be possible to disentangle one aspect of a laboratory, say its funding, from another, the knowledge that is produced there, it would be misleading to do this. These and all the other component parts of a laboratory have effects on other parts. In these essays specific aspects of laboratories are studied, but with due attention to how they are affected by and affect other parts of laboratory life. All these approaches have their place and should be viewed as complementing one another. In the absence of a detailed laboratory typology and of studies based on research of more than a handful of laboratories, then no approach can or should be said to be superior to another. But taken together one form of approach may serve to illuminate another, as is illustrated by Dominique Pestre's chapter.

These particular studies have been brought together to allow historians, sociologists, and, indeed, all those interested in laboratories, to explore the possible methods for understanding the workings of this architectural space which is unique to our scientifically and technologically orientated culture.

1 The Origins and Early Evolution of Laboratories

It is almost impossible to conceive of chemistry outside a laboratory environment; the dictionary definition particularly applies the word to chemistry. It was in this discipline that laboratories were first built and perfected as places where nature could be studied under controlled conditions. June Fullmer's and David Fenby's chapters illustrate the close interaction that existed between chemical research and teaching institutions and laboratories in the eighteenth and early nineteenth centuries. Of course these laboratories developed out of the private laboratories that had existed previously. But in the eighteenth century the institutional need for chemical laboratories for teaching or for research came to be recognized. Even so, the desire for private laboratories still existed, as Brian Gee's chapter on portable laboratories in the nineteenth century indicates.

It took most of the nineteenth century before physics followed chemistry in becoming a laboratory subject based in institutions. Despite the example set by the Royal Institution, very few institutions had laboratories in which physics experiments could be conducted; even the Royal Institution laboratories remained predominantly chemical for most of the nineteenth century. Much experimental physics at this time was done in private laboratories. One has only to think of Andrew Crosse's electrical laboratory in his Somerset stately home[2] or G. G. Stokes discovering fluorescence in his laboratory in his home in Cambridge[3] to realize the limited number of institutions which possessed physics laboratories.

In the latter part of the nineteenth century this altered and more institutions began to have physics and other laboratories. At Harvard (discussed by Aronovitch) and Cambridge (discussed by Falconer) physics laboratories were founded. The subjects which became suitable for laboratory study also increased in number. For example, David Cahan's chapter discusses the precise measurement of physical quantities in a laboratory, while W. H. Brock examines the laboratory teaching of engineering.

But perhaps it is in this century that laboratories have enjoyed their greatest success. Andrew Pickering shows how the success of laboratories during World War II was a significant contribution in deciding to undertake a programme of building large particle physics laboratories in the United States such as Fermilab, discussed by Westfall. Laboratories have now become so large that some require multi-national funding, as at CERN, discussed by Krige and by Pestre.

4 INTRODUCTION

2 Funding and Building of Laboratories

The development of laboratories to form ever larger and encompassing ever more scientific and indeed non-scientific subjects is a complex subject. Each essay comments, either at length or in passing, on the need for money to support laboratories. June Fullmer looks at the way Humphry Davy raised money in the first decade of the nineteenth century to expand his electro-chemical research, based at the newly founded Royal Institution, in the belief that larger electric cells would produce results even more spectacular than his already-made discoveries of sodium and of potassium. He was able to use his experience in fund-raising gained while he was with Thomas Beddoes in Bristol. Davy had learnt to deploy the patriotic and religious rhetoric of his time to manipulate and secure the interests and patronage of the people who had money. He was thus able to successfully raise the necessary funds for his research and in the process transform the Royal Institution from a teaching and consultancy institute to one where research came to play an ever larger role.

American physicists in the 1950s and 1960s argued that if larger particle accelerators were built, then even more important discoveries would be made in particle physics (an argument not too dissimilar to one of Davy's). To obtain sufficient funding for what would be large accelerators, physicists, as Andrew Pickering argues, had to match military interests with their own. Pickering does not discuss laboratories but he does provide an analysis of the way physicists were able to obtain the patronage of the military establishment. One of the interests of this establishment was the need to maintain a large number of highly scientifically trained personnel who could be used in times of crisis. This matched with the interests of leading physicists in developing large projects and, hence, large laboratories. It is worth remembering that research laboratories produce trained scientists as well as knowledge. The dovetailing of interests was translated into the provision of vast sums of United States Government money to build a number of particle accelerators. How some of this money was spent is discussed in Catherine Westfall's chapter on the foundation of Fermilab. Here, in addition to the general context outlined by Pickering, problems of conflicting interests within the scientific community and within United States national and regional politics came into play to eventually determine the midwest location of the laboratory.

While patronage plays a large and dominant role in deciding the location of laboratories, environmental factors play an important, and in some cases decisive, role. The locations of the laboratories discussed above were already fixed either owing to the presence of existing buildings or by the eventual resolution of conflicts of interest. Once a location had been fixed, then the laboratory was built to take best advantage of the site.

Mari Williams's chapter on the Pulkowa observatory provides an example of the relationship between patronage and location. The determinants for the observatory site involved clear skies, an uncluttered southern horizon and access to the patronage of the court at St. Petersburg, which all contrived to make Pulkowa the best site. The observatory that was built allowed Struve and his team to become a community centred on astronomical observations. The initial choice of the site allowed it to function effectively until the 1940s. Half a century or so later, the problems of location and development which faced the Physikalisch-Technische Reichsanstalt, discussed in David Cahan's essay, were much more complex. The Physikalisch-Technische Reichsanstalt needed to be near Berlin and the government, and so, thanks to Siemens' additional private patronage, the site at Charlottenburg was chosen. But very soon problems over isolating the Physikalisch-Technische Reichsanstalt from its external environment became acute. Mechanical vibrations and electromagnetic disturbance from trams seriously interfered with the operation of the apparatus. These problems were eventually, but not easily, resolved by negotiation with the tram company. Both Pulkowa and the Physikalisch-Technische Reichsanstalt were new laboratories and so the location of the sites and the solution of the environmental problems were matters for negotiation. In the case of Harvard Physics laboratory, discussed by Lawrence Aronovitch, the location of this was already fixed, but the site itself encouraged patronage. The technical problems of the environment, particularly shielding apparatus from vibrations and from electromagnetic disturbance had to be overcome, as at the near-contemporary Physikalisch-Technische Reichsanstalt; with, it might be added, somewhat less success.

These essays discuss the establishment of specific laboratories in Britain, Russia, Germany and the United States. In these large industrialized and industrializing countries a comprehensive survey of the development of laboratories has yet to be undertaken. Such a survey, for these countries, at least, would require a considerable social research programme which, while it would be invaluable, is hardly likely to be undertaken in the near future.

3 Change in Established Laboratories

Like any human institution, laboratories are subject to change. Unlike their foundation, when laboratories are influenced very strongly by the interests of their patrons, their development is much more linked to internal factors. The fact that a laboratory is materially present affects how people think and support it. There is not much else one can do with an established laboratory apart from some sort of science — although the content does of course change over time.

The nature of this change for a particular laboratory is dependent on local circumstances. For example, at the Cavendish Laboratory in the University of Cambridge, discussed by Isobel Falconer, the type of research that was pursued under two successive professors, Lord Rayleigh and J. J. Thomson, changed dramatically. The very precise Tripos-orientated type of research conducted by Rayleigh passed to the qualitative experimental hypothetical (one might almost say speculative) research carried out by Thomson and his students. This attracted a large number of students to the Cavendish who subsequently became leading physicists in their own right. It is worth recalling that, as with particle physics, research laboratories produced trained scientists as well as knowledge. In this case a single individual was able to change the research direction of an established laboratory. This could only happen because of lack of resources; had the Cavendish possessed large resources in terms of apparatus, that, in itself, would direct the course that research took irrespective of the professor's interests at a given moment.

This was the situation which confronted CERN. While on the one hand a large quantity of money had been put into establishing what was at the time the world's largest proton synchrotron, on the other hand comparatively little had been invested in purchasing the equipment necessary for the conduct of research. The consequence of this was that American accelerators discovered new fundamental particles with less powerful accelerators than CERN; this is detailed in John Krige's essay. What this illustrates is the difficulty CERN had in making the transition from building machines to exploiting them adequately. To change the activities of a large laboratory takes more time, money and managerial skills than at smaller ones.

Managerial problems in accelerators can in theory be overcome by planning ahead. This is the subject of Pestre's essay, also on CERN. The first debate at CERN concerning the future of its research programme was undertaken even before the machine had worked and long before it gave its first important results. In this debate it was the interests of the physicists and the builders which had to be reconciled. At this point there were no problems for the patrons (various European governments) engaging in these discussions. Once an academic laboratory has been established, the active role of the patrons in deciding what happens in the laboratory declines sharply. This, however, is not the case in technological laboratories.

4 The Production and Dissemination of Knowledge through Laboratories

In the case of large particle accelerators, the knowledge that is produced using them is highly structured by the way they were designed, what they are capable of, and so on; they require recording equipment as an integral part. In laboratories such as the Cavendish at the time discussed by Falconer, experiments were done as inexpensively as possible by individual researchers.

David Gooding's chapter reconstructs Faraday's experimental procedures from his laboratory notebook. Faraday, working in the research laboratory that Davy had founded, discovered electromagnetic rotations in September 1821. Analysing the text of Faraday's notebook entries for that time in the context of a physical reconstruction of the experiments that Faraday conducted, Gooding provides a detailed account of this work in which the difficulties — tacit knowledge and experience that goes unrecorded — of notebook accounts are brought out.

Faraday's work was of strictly scientific interest, conducted in a laboratory that was supported by chemical consultancy work. In fact, there were at that time hardly any strictly academic laboratories in existence and few industrial laboratories. Yet less than a century later there were many laboratories of different varieties spread throughout the industrial world. The type of patronage for different laboratories structured different approaches to knowledge.

Also, as noted above, research laboratories produce trained scientists as well as knowledge. In these there is a close relationship between research and training. But there are also specialist teaching laboratories. Knowledge that has already been established can be packaged in many forms. The place of portable laboratories and amusement chests, discussed by Brian Gee, exemplifies one way in which that knowledge can readily be assimilated through commercial interest. Here the interests of the trade, through its instrument makers and chemical apparatus suppliers, were particularly influential intermediaries between the scientific community and the wider public. They were able to digest and simplify experiments and demonstrations and so make provision for a practical experience among those for whom formal education, whether at university, school or mechanics' institute, had yet little or no laboratory facility.

More conventionally at Glasgow University, chemistry was taught from an early point to students who took part in practical laboratory work. This, as discussed by David Fenby, was the outcome of the tradition of using demonstrations to teach chemistry. (Itself a consequence of chemistry teaching being associated with medicine, where this form of teaching was common.) Related to technological interests was the foundation of the Finsbury Technical College, which, as W. H. Brock shows, was specifically designed with practical laboratories to provide a technical education. Thus, during the nineteenth century the purpose of the laboratory expanded both in the areas of knowledge which could be studied there and in the formalization of that knowledge for pedagogic purposes to serve the ever increasing technical industrialization of society.

Notes and References

(See Bibliographical Note on page 242 for use of notes)

1. Exceptions that should be mentioned here are Lartour and Woolgar (1979) and Knorr-Cetina (1981).
2. Secord (1989).
3. Larmor (1907), Vol. 1, pp. 8–10.

Section 1

CHEMICAL LABORATORIES

1

Humphry Davy: Fund Raiser

June Z. Fullmer

In all scientific settings one equation operates with the universal applicability of Newton's Laws: laboratories equal outlay of money. With the exception of Pythagoras, who, according to poetic myth, traced his demonstrations in sand with a twig, it is hard to think of any scientist who has never been forced to raise funds to support continuing research. At the beginning of the nineteenth century, fund-raising became an acute problem. An expanding scientific and technological enterprise generated its own needs for further expansion; hindsight reveals embryonic features of what we now call 'Big Science'. However, traditional sources of patronage had perceptibly weakened or vanished.[1] New routes for financial support of research had to be plotted.

Humphry Davy (1778–1829) trod some of these poorly marked paths, for on three separate occasions he had to take part in fund-raising to support his laboratory research. These instances show how diverse fund-raising attempts could be at the beginning of the nineteenth century, how they were tailored by their proponents to meet specific needs and how they were made to fit their institutionalized setting.

1799 dates Davy's first involvement with fund-raising, when he was at Bristol working at Dr Thomas Beddoes's Pneumatic Institute. His second and third tries were made at the Royal Institution in 1808 and again in 1810. An *ad hoc* air pervades all three attempts. In the absence of established procedures, scientific researchers relied on their own fund-raising ingenuity to pay for their laboratories and for their laboratory equipment. What methods they attempted depended on their views of the nature of laboratories and of what sort of science was to be carried out in them, as well as on the circumstances in which the scientists found themselves.

Analysis of the personality, aims and methods of Davy's original employer helps to explain how he was first drawn into fund-raising. When Beddoes (1760–1808) settled in Bristol, he could already look back upon a career marked by accomplishment and by controversy.[2] In 1786 he earned an Oxford M.D.; his medical education also included training at Edinburgh. At Bristol he sought cures for tuberculosis and the diseases he thought related to it, but his interests encompassed more than medicine. Wherever he was he espoused causes that he thought might alleviate the ills of the human condition, be they political, social or medical. For example, angered by what he saw as Pitt's failure to remove poverty with its attendant social evils, he lashed out against him in a famous pamphlet, 'Essay on the Public Merits of Mr. Pitt'.[3] Moreover, Beddoes's knowledge of chemistry was extensive, despite his having little interest in or talent for laboratory work.[4] As soon as he had learned of Lavoisier's contributions, he became one of the earliest British public advocates of Lavoisier's tenets. Before coming to Bristol Beddoes had taught chemistry at Oxford, spreading his Lavoisier-based chemical doctrines to all who would listen. In a certain sense the Pneumatic Institute he created represented a union of his medical and chemical beliefs with his social concerns; he hoped to find a way to treat many diseases through the chemical powers he thought gaseous inhalations might provide, thereby alleviating some of the suffering of humankind.

There can be no doubt that Beddoes played a major role in Davy's life. Not only did he teach him much about chemistry and physiology, but also he opened the scientific literature to him, and he infected him with a (misguided) enthusiasm for a certain brand of metaphysics. Above all, Beddoes taught him how to behave as a research scientist. Davy, always a quick student, learned his lessons so thoroughly that it is possible to see him following some of Beddoes's precepts throughout his professional life.

One important lesson Davy imbibed from Beddoes centred on the problem of raising funds for laboratory experiments. Even in 1798 laboratories did not come cheap. From his first published papers in Contributions to Physical and Medical Knowledge, Principally from the West of England, Collected by Thomas Beddoes, M.D.,[5] it is obvious that at the Pneumatic Institute Davy had access to a goodly variety of chemicals and laboratory apparatus, as well as access to most of the recently published scientific journals and books. He worked in a setting especially designed for the needs of the Institute; 'above all', Davy wrote to his mother, 'I have an excellent laboratory'.[6] The entire operation represented a considerable outlay of 'ready money'.

Beddoes's pursuits stamp him as unique among medical practitioners. The late economist Joseph A. Schumpeter isolated five characteristics marking a technological innovator–entrepreneur.[7] Inspection of that list shows that Beddoes, while not ordinarily classed as a 'technologist', certainly fits all of Schumpeter's criteria: first, Beddoes formed a 'new firm' which did not arise from an older one, but started working in competition with 'firms' (i.e. other medical practitioners) that used a standard technology;[8] second, Beddoes can be characterized as having 'initiative' and 'authority';[9] third, Beddoes had to convince his financial backers of the validity of his proposals;[10] fourth, Beddoes retired from the arena of medical innovation only when his strength was spent;[11] and, finally, Beddoes was a man driven by a dream.[12] He was a medical entrepreneur–innovator on a scale that proved adequate to the needs of his Institute.

Schumpeter's analysis has been extended by F. M. Scherer through his introduction of the concept of the 'developer'.[13] He used as his primary exemplar the activities of two of Beddoes's associates, Matthew Boulton and James Watt, when they created and marketed their steam engines. In the light of Scherer's model, Davy in Bristol can be said to have functioned as 'medical-developer', that is to say, as the individual chiefly responsible for transforming Beddoes's 'invention', the treatment of illness through gaseous inhalation, into a feasible, innovative medical procedure. This role accounts for Davy's initial involvement as a laboratory fund-raiser.

Beddoes's intention of founding an Institute for the practice of pneumatic medical treatments first found voice in 1791, when he was visited by his friend of long standing, Davies Giddy (1767–1839) in Cornwall;[14] his ostensible aim on the trip had been to study the geology of the area, but while he and Giddy were together they discussed Beddoes's grand project. By 1793 he was actively seeking a site for his Pneumatic Institute. Although he relied primarily on personal connections to fund his dream, raising money for the operation took him longer than it had taken him to find a suitable location in Clifton, a suburb of Bristol. His first backers were three of his old friends, each of whom contributed £200. In 1795, a month before he died, Josiah Wedgwood contributed £1000.[15] In 1796 a published list of subscribers to Beddoes's Institute included many of the other prominent members of the Lunar Society (to which Beddoes belonged) together with their families.

Despite this generous backing, Beddoes vigorously continued his fund-raising efforts. Necessity, rather than greed,[16] forced him to a variety of stratagems to keep the money for his enterprise flowing. Davy was drawn into one of them, for Beddoes chose to hope that publication of the *Contributions to Physical and Medical Knowledge* . . . might become a source for funds. He announced it as the first of a series of works; proceeds arising from the publication of each volume in the projected series were to go to support some medical charity. Perforce the first volume would support the Pneumatic Institute, but subsequent beneficiaries were to be determined by lot 'from the benevolent establishments of the counties comprised in the circuit'.[17] Nine-tenths of the volume (205 pages) were written by Davy; a folded plate included in it depended on Davy's newly minted chemical nomenclature as well. Thus Davy, willy-nilly, was to attempt to support his own efforts from earnings produced partly by his writing.

Beddoes, like all entrepreneurs, was a bold risk-taker.[18] The ingenuity of his scheme to raise money for a group of hospitals or infirmaries through publication is impressive: the results were less so. Only the first volume of the projected series appeared. The proposed lotteries, founded, as are all lotteries, on promises and hopes, could only be conceived in the imagination of a sanguine, dream-driven entrepreneur–innovator. Davy, too well acquainted with entrepreneurial risk,[19] was no such gambler; he confined his risk-taking to intellectual play with theories about the natural world and the ways in which it operated. Still, the lesson Beddoes taught him emerged clearly: a researcher is required to help raise the funds to pay for his laboratory.

When Davy came to the Royal Institution in 1801, he found himself in a quite different milieu. Certainly money was needed to support experiments and applied research, but entrepreneurial innovation of the kind exercised by Beddoes was distinctly out of place. The founders of the Royal Institution had a charter from the Crown; moreover, the Institution had not only the blessings but also the guidance of Sir Joseph Banks, the scientific panjandrum of the age. Now Davy had become part of an organization so constituted that fund-raising could be handled along institutional lines.

In the first several years following his arrival the Royal Institution suffered financial ups-and-downs.[20] Davy as lecturer–demonstrator helped to attract large numbers of annual subscribers, including well-to-do women, and, thus, more money for operating expenses. In 1807 his discovery through the use of the voltaic pile of the new elements sodium and potassium, followed by his announcement in 1808 of the isolation of calcium, magnesium, strontium and barium, and other elements,[21] simultaneously piqued and gratified his Royal Institution audiences. Davy told them that 'the voltaic battery was an alarm-bell to experimenters in every part of Europe', as indeed it was.[22] Examination of the literature shows that the tocsin sounded in Munich, in Jena, in Kiel, in Regensburg, in Geneva, in Moscow, in Vienna, in Erfurt, in Verona, in Chemnitz and in Stockholm, but that it rang especially loudly in Paris.[23] Davy's revelation of a cornucopia of new elements made it appear as if only the size of the available electrochemical piles prohibited more startling discoveries.[24] An international race was on for bigger and bigger batteries; although Davy did not mention France as his principal competitor, news of Napoleon's generous grant to the École Polytechnique for construction of a large voltaic battery had filtered to England. The two nations were already at war. To be bested on the scientific front was not especially appealing.

To raise the necessary funds to build a new, large pile, Davy proceeded in a fashion that now seems almost orthodox. He first applied to the Board of Managers of the Royal Institution for permission to build it and to raise the sum required for its construction. At their meeting of 11 July (the Managers present were Lord Dundas, W. Watson, Thomas Bernard and Charles Hatchett) he laid before them the following proposal:

A new path of Discovery having been opened in the agencies of the Electrical Battery of Volta, which promises to lead to the greatest improvements in Chemistry and Natural philosophy and the useful Arts connected with them; and since the increase of the size of the Apparatus is absolutely necessary for pursuing it to its full extent, it is proposed to raise a Fund by subscription for constructing a powerful Battery, worthy of a National Establishment and capable of promoting the great objects of science.

Already in other Countries public and ample means have been provided for pursuing these investigations. They have had their origins in this Country, and it would be dishonourable to a nation so great, so powerful, and so rich, if, from the want of pecuniary resources, they should be completed abroad.

An appeal to enlightened individuals on this Subject can scarcely be made in vain. It is proposed that the Instrument and apparatus be erected in the Laboratory of the Royal Institution, where it shall be employed in the advancement of this new department of Science.[25]

The Managers immediately endorsed the proposal, agreed to subscribe and caused a book to be 'opened in the steward's office for the purpose of entering the names of all those who ... wish to contribute towards this important national object'.[26] Davy in his lecture series made a public appeal for additional funds; donations amounted to £520.

When a road bends to a new direction, we cannot point to a precise turning point; even after we have travelled the road, all we can say is that its direction has indeed changed. Historical hindsight operates in much the same way. This instance permits us to say that Davy's move proved decisive, not only for his own work, but also for the funding of laboratories. The Royal Institution had not been founded as a research institute; the original proposal announced:

The two great objects of the Institution [are] ... the general diffusion of the knowledge of all new improvements, in whatever quarter of the world they may originate; and teaching the application of scientific discoveries to the improvement of arts and manufactures in this country, and to the increase of domestic comfort and convenience. . . .[27]

The new research mission of the Royal Institution and its research eminence derived more from Davy's 1808 fund-raising than from the aims of the earliest founders, of which Count Rumford, Sir Joseph Banks and Thomas Bernard, among others, were the principals. The 1808 funding effort relied on patriotism and on fear of French supremacy[28] for its chief appeal, but Davy's subsequent comments showed that, although he had not abandoned nationalistic arguments, for him more than national pride was at stake.

When in his public lecture he thanked the donors of the £520, he acknowledged the shift of direction in the road he was travelling. In addition, he plainly demonstrated how the management of the Royal Institution then viewed research. Support of scientific research by funding laboratories, he said, should be seen as an act of charity: laboratories were to be regarded as a fitting object for philanthropy:

A scientific institution ought no more to be made an object of profit than an hospital, or a charitable establishment. . . . What this Institution has done . . . has tended to the progress of knowledge and invention. . . . With more ample support, more, undoubtedly, would be effected. With a devotion to the experimental sciences and arts, nothing but good could result from an extension of the undertaking: and it is no mean object to the country, that *the first attempt of this kind* should succeed.[29]

Further in this lecture Davy promised his supporters a new kind of compensation: their 'first attempt' at scientific philanthropy would produce important discoveries.

Study of the early days of the Royal Institution laboratories has produced at least two extreme interpretations. Some scholars have viewed the laboratories as the cupel which refined to purest metal what were to become Victorian ideals of scientific research.[30] In another view, those early days were a crucible for licence; landed agriculturists and nabobs alike sought and sometimes found in the laboratories ways to extend their power and wealth.[31] Despite their differences, these interpretations share two tenets. Both criticized Davy for his apparent failure to do theoretical research;[32] both recognized that numbered among the early founders of the Royal Institution were some of Britain's leading philanthropists.[33] Morris Berman, especially, saw the Royal Institution as a vehicle for the administration of 'scientific philanthropy', i.e. philanthropy along lines defined by Jeremy Bentham.

Criticism of Davy because his theoretical publications seem deficient probably stems from Whiggish tendencies inherent both in Victorian attitudes to science and in many, more recent Marxist appraisals. Unless we steel ourselves, we can easily slip into the same sort of criticism; in some circles there is a tacit conviction that theoretical work is more precious than experimental work. In Davy's defence it should be pointed out that, while he was first and foremost an experimentalist, his success as an experimenter did not come by happenstance; each of his experiments was 'theory-loaded'. William Nicholson (1753–1815), whose *Dictionary of Chemistry* Davy had conscientiously studied when he taught himself chemistry, enjoined every aspiring chemist to strive for a happy marriage of theory with experiment.[34] Here, too, Davy had proved himself an apt pupil. Moreover, the places and the time in which he found himself valued experiments above all. In Albemarle Street Davy saw to it that theories were of interest not for their own sake — not all of his contemporaries shared his view — but for their interplay with what went on in the laboratory.

It is no mystery why the support of research laboratories became for a while an object for charity, just as hospitals were, but it is a slight mystery why this state of affairs has been so often underemphasized.[35] Victorians, latter-day Victorians, Edwardians and Marxists alike agree that the founders of the Royal Institution included the era's leading philanthropists; the Earl of Winchelsea, first President of the Managers, and Thomas Bernard (1750–1818), first Secretary of the Royal Institution, are premier examples.

Once at the Royal Institution, Davy experienced their palpable influence. In particular, Davy formed an alliance with Thomas Bernard. In September of 1805, while the two were touring companions, Davy wrote to his early sponsor, Davies Giddy:

> Much kindness and long knowledge of him [Bernard], have made me partial to that gentleman, and may perhaps influence me when I say, that there is not a more patriotic, good, and public-spirited man in Great Britain.[36]

Their admiration was reciprocal. Bernard very much hoped that Davy might be convinced to enter the church; as an inducement he promised him a good preferment. Davy's tactful rejection of Bernard's offer permitted him to say something about his own philosophy and about his aims in lecturing at the Royal Institution. In addition, his statement offers clues about the level of discourse between the two. After thanking Bernard for his interest in his 'public labours', Davy wrote:

> I am never more delighted than when I am able to deduce any moral and religious conclusions from philosophical truths. Science is valuable for many reasons: but there is nothing that gives it so high and dignified a character, as the means which it affords of interpreting the works of nature, so as to unfold the wisdom and glory of the Creator. Be assured, my dear Sir, that I shall lose no opportunity of making those deductions which awaken devotional feelings, and connect the natural with the moral sense. And I hope my claims to your approbation, and to the approbation of men who, like you, combine pious sentiments with noble and enlightened views, will not diminish. . . .[37]

If Davy had found in Beddoes a role model of how to behave as a professional scientist, in Bernard he found a role model of another sort.

Although Bernard was a solid Churchman, he differed from his Evangelical philanthropic confreres in several important ways; his attitude towards philanthropy, his industry and his powers of persuasion for the causes he championed simultaneously made him the most outstanding philanthropist of his time and unique. He was the son of Sir Francis Bernard, the last Royal Governor of Massachusetts Bay Colony, educated first at Harvard and then at Middle Temple. Fifteen lucrative years as a conveyancer coupled with a financially rewarding marriage led him to retire and to pursue philanthropy:

> When I thought I had acquired in my Profession such a competence as satisfied my desires, I determined to quit the Law, & try what useful Occupation I could find that was not likely to increase *l'embarras des richesses*. The Endeavour to meliorate the domestic habits of the labouring Class, was the first amusement that occurred.[38]

Bernard thought that philanthropy offered the titillation 'of the gaming Table without its Horrors', coupled with the satisfaction of productive accomplishment. Evidently he shared some of Beddoes's risk-taking proclivities, but he chose to express them differently.

Bernard's name will be forever associated with the Society for Bettering the Conditions and Increasing the Comforts of the Poor — known familiarly as 'The Bettering Society' — as well as with the Royal Institution. The web of connections between the two establishments was intricate: at the end of the eighteenth century Bernard's enthusiasm for Count Rumford's schemes knew no bounds;[39] as a result, before the first organizational meeting creating the Royal Institution, Rumford's plan for it had been approved by The Bettering

Society. Although Rumford soon lost his zeal and was encouraged to allow his ties with the Royal Institution to languish, Bernard continued to work for it until his death.

Nonetheless, the support of scientific laboratories as a charity which Davy invoked in 1808 differed in one important way from other early nineteenth century philanthropies with which Bernard was associated. All of them had been directed to ameliorating the conditions of individuals in need. Now support for laboratories aimed at facilitating a wider undertaking — improvement of the quality of life for everyone.

Undeniably Davy in his lectures and in his appeals for funds first mirrored Bernard's philanthropic inspiration. Later he nimbly adapted that inspiration to meet a new situation. As he did so, he aired the crucial alteration that had occurred in the way in which laboratories, experiments and research had come to be regarded at the Royal Institution. Even as the Royal Institution became 'unRumfordized', it rapidly transformed to become 'Davyized'.

Nowhere is that change more visible than in Davy's 1810 explanation for the revisions which would modify its financial base. These modifications have been much discussed[40] and it is plain that Davy was not the sole architect of the new design. His overt role in the business was as publicist. Yet it must also be recognized that to a great extent Davy did influence the final outcome. To place the Royal Institution on a sounder financial base meant, first, expanding that base. A direct route to expansion lay in democratizing the leadership.[41] That democratization brought with it, however, a need to redefine the role of the research laboratories. When Davy delivered his lecture about the reorganization, his special point of view on the nature of the scientific enterprise had prevailed.

Appraisals of his lecture vary. Bence Jones found it 'a reflection of Davy in his full power';[42] in 1916 the Managers of the Royal Institution thought enough of the lecture to reprint it in *The Proceedings*;[43] most recently Berman found it both vague and equivocating, and, finally, a not very convincing performance.[44] Extrinsic merits aside, the lecture is a model of contemporary philanthropic entreaty — a form of which Bernard was master, and a form at which Davy proved adept — with the added fillip that in this instance the entreaty was for an undertaking which had itself recently undergone change.

Davy's organization of the lecture was strategic. He adopted classical rhetoric's three modes of persuasion: he appealed to his audience's reason (*logos*); he appealed to their emotions (*pathos*); and he modestly appealed to them through his own authority (*ethos*). He began by defining the Royal Institution through a brief recapitulation of its history and early aims. He reviewed the reasons why the Model collection had failed — it could give away manufacturers' secrets — but then he pointed out that the 'Laboratory has always been open for carrying on any investigations connected with the progress of chemical science or the arts. It has afforded the means of promoting inquiries of some importance to the community.'

This statement of aims contrasts sharply with what Davy saw as the advantages to be derived from the new plan. While the 'open laboratory' had been of 'some importance to the community', the reorganized Royal Institution laboratories would now become part of a larger vision:

> . . . let us not neglect that basis on which the greatness of modern times, and of our own country, so peculiarly rests; Experimental Philosophy and the Experimental Arts. Let their merits be justly estimated, and set forth with dignity and truth; let not the countrymen of Bacon, of Newton, and of Boyle, neglect those pure springs of knowledge from which those great men drew such copious supplies, both for profit and for glory: and let it not be forgotten, that Science has its moral and intellectual, as well as its common uses; that its object is not only to apply the different substances in Nature, for the advantages, comfort, and benefit of man, but likewise to set forth that wonderful and magnificent History of Wisdom and Intelligence, which is written in legible characters, both in the Heavens and on the Earth.[45]

Davy's lofty peroration could have left no doubts in the minds of his hearers: they would be involved in the philanthropic support of laboratories. Support of those laboratories would both glorify the Deity and serve the accretion of knowledge for its own sake.

In one sense it is possible to say that Davy had come full circle. The injustices inflicted by poverty, the crimes perpetrated by restricted liberty and by inadequate provisions for the poor which had so moved Beddoes, could possibly now be tempered by the ministration of scientific discoveries. Still, a new spirit obtained. When the Managers, Visitors and Proprietors at meeting after meeting discussed how best to reorganize the Royal Institution, Davy, by 1809 one of the most renowned of the British scientific practitioners,[46] sat with them. He was unstinting in his praise of philanthropic aims, but he acclaimed a philanthropy with a difference. Were suitable reforms adopted, were the Institution put on a stable financial base, neither Davy nor his successors should again have to plead a charitable case for the support of scientific laboratories. Rewards awaited those who would support a research laboratory. The national interest would be preserved, the traditions of Bacon, Boyle and Newton honoured by the new practices. The laboratory — his enthymeme is apparent — given adequate financial support, would contribute to the moral and intellectual stature of the donors.

Finally, in his lecture Davy celebrated almost in the same breath both Divine Wisdom and the results of laboratory research. He had kept his earlier promise to Thomas Bernard. But he did more than that. He succeeded in converting the laboratory almost into a place of worship; if before it had been a charity ward, now it had become a temple. His tripartite appeal to religious impulse, to nationalistic pride and to charitable inclination permitted Davy to argue for laboratory research for its own sake. It should not have been a surprise that what Davy called for in his lecture, and what he had promoted throughout the long campaign for restructuring, was a laboratory created along lines tailored to his own beliefs and practices.

The new plan permitted to a certain extent democratizing the governance of the Royal Institution. As a result, the argument ran, output from the laboratories should benefit far more people than it had in the past. In the equation a laboratory equals outlay of money, Davy had publicly called attention to its fundamental discriminant: the way laboratory findings will be used is a function of the source of the money that pays for them.

Notes and References

1. Brock (1976).
2. For biographical details see Stock (1811). See also Stansfield (1984).
3. Poynter (1969), p. 63n, found Beddoes 'one of Pitt's most severe critics'; Beddoes 'thought the Prime Minister's philanthropy a sham and his abilities mediocre' (p. 66).
4. Late in his life Davy recollected that Beddoes 'was little enlightened by experiment & I may say little attentive to it'. Fullmer (1967), p. 131. Davy's remark shows that he himself found experimental results both illuminating and instructive.
5. Davy, H. (1799a, b).
6. Davy to his mother, Mrs Grace Davy, 11 October 1798; the letter is in the Archives of the Royal Institution.
7. Schumpeter (1934). The work has recently been discussed by Staudenmaier (1985), p. 56.
8. Schumpeter (1934), p. 16.
9. *Ibid.*, pp. 75 and 78.
10. *Ibid.*, p. 89.
11. *Ibid.*, pp. 78 and 92.
12. *Ibid.*, p. 91.
13. Scherer (1965). Staudenmaier (1985) discussed the same work on pp. 56–8.
14. Todd (1967), pp. 25–6.
15. Wedgwood contributed the money with the hope that Beddoes would demonstrate that gases 'would not be serviceable to medicine'. Schofield (1963), p. 374.
16. He may have been greedy for food, since he was monstrously fat for much of his life. His detractors were quick to point it out: 'he is so fat and short that he might almost do for a shew', one of them said. Stock (1811), p. 36.
17. *Ibid.*, p. 157.
18. Beddoes manifested this trait in several ways. Thus, from the time he was an undergraduate he had a reputation as a formidable whist player. Stock (1811), p. 10, says 'he was supposed to play that game as well as almost any man in England'.
19. When Davy's father, Robert Davy, died in 1794 (Davy had just turned 16), his mother was saddled with debt largely incurred by Robert Davy's ventures in Cornish mines.
20. How financially troubled the Royal Institution was has been the subject for argument. See Bence Jones (1871), Appendix III, p. 425 *et passim*. Vernon (1963), maintained that the Institution's financial solvency was directly related to Davy's ability to attract audiences. Berman (1978), pp. 129–30, argued that the Institution was not financially stable until well into the Faraday years.
21. His discoveries were announced to the Royal Society in Bakerian lectures and published in the prestigious *Philosophical Transactions* — a far cry from the single volume of Beddoes (1799).
22. Davy made the statement in a lecture delivered in 1810. Davy, J. (1839–40), Vol. 8, p. 271.

23. The list is based on an examination of the 1808 volumes of the following journals: *Ann. Chim.*, **65**, **66**, **67**, **68**; *Ann. Physik (Halle)*, **28**, **30**; *Bull. Sci. Soc. Philomathique*, **1**; *J. Chem. Physik Mineral.*, **5**, **7**; *J. Phys. Chim. Histoire Nat. Arts*, **66**, **67**; *Bibl. Brit.*, **37**, **38**, **39**.

24. A rage for larger and larger piles gripped many workers. In England John George Children, a friend of Davy's and a member of the Royal Institution, built a pile so large that it had to be set up out-of-doors, with winches to handle the giant plates. Children (1809–15).

25. Also quoted by Bence Jones (1871), pp. 355–6.

26. Paris (1831), Vol. 1, p. 316. See also Bence Jones (1871), p. 356.

27. *Ibid.*, p. 121.

28. Fear as a motive for philanthropy has been much discussed. See, e.g., *Report of the Princeton Conference on the History of Philanthropy in the United States*, Russell Sage Foundation, New York, 1956, p. 11. I am grateful to Professor Robert Bremner for pointing out the reference.

29. Davy, J. (1839–40), Vol. 8, pp. 357–8. Emphasis added. Davy was correct in his assessment that this represented the 'first time' support had been arranged for a piece of laboratory apparatus by public subscription. To be sure, in 1806 the Royal Institution raised £942 by public subscription for the establishment of a Mineralogical Collection and an Assay Office. While Davy subsequently himself performed many of the assays, he obviously did not regard the operation as part of the laboratory of the Royal Institution, as his comments here indicate. Berman (1978).

30. Bence Jones (1871), *passim*. See also Thorpe (1896) and Hartley (1966).

31. See Berman (1978), *passim*.

32. Siegfried (1959); Hartley (1960); Berman (1978), pp. 71–4.

33. Bence Jones (1871), Chap. III *et passim*; Berman (1978), pp. 1–31 *et passim*.

34. 'The enlightened cultivator of Science will be neither of these [i.e. neither a theoretician nor a 'man of fact'] exclusively, or rather he will be both.' Nicholson (1795), Vol. 1, pp. v–vi.

35. This paper, concentrating as it does on Davy as fund-raiser, ignores the influence of the agricultural interests on the early Royal Institution. Thomas Bernard, in addition to being a philanthropist, was also a prominent land-owner with a strong interest in agricultural matters. Berman (1978), Chap. 2.

36. Paris (1831), Vol. 1, p. 204.

37. Davy, J. (1836), Vol. 1, pp. 272–3.

38. Quoted by Owen (1964), p. 105.

39. Redlich (1971), p. 202.

40. Bence Jones (1871), pp. 290–302; Berman (1978), pp. 96–9.

41. *Ibid.*, pp. 95–7.

42. Bence Jones (1871), p. 292.

43. Davy, H. (1810). The *Annual Register* (1809), **51**, pp. 822–6, contained 'Sketch of a Plan for Improving the Royal Institution, and Erecting it on a Permanent Foundation' which was probably also written by Davy.

44. Berman (1978), pp. 97–8.

45. Davy, H. (1810), p. 16.

46. In 1810 Davy's English reputation was chiefly confined to that part of the population best described as 'scientific cognoscenti'. 'Philalethes, Jr.', who wrote to Tilloch from Chichester, *Phil. Mag.* (1808), **32**, pp. 62–6, displays the regard in which he was held: 'I say, Mr. Davy's discoveries are *glorious* — for they were not the result of a parcel of *guesses*; but . . . of a fine train of reasoning, from data of *his own*. I am at a loss which to admire most, the penetration of his genius, or his persevering and accuracy in prosecuting his researches.' Davy's reputation with those who were not especially knowledgeable in scientific matters emerged after his invention of the miners' safety lamp.

2

The Lectureship in Chemistry and the Chemical Laboratory, University of Glasgow, 1747–1818

David V. Fenby

The standard method of teaching chemistry, from its emergence as a university discipline in the seventeenth century until the early nineteenth century, was the lecture accompanied by lecture demonstrations. This approach was related to changes that had taken place in medical education. At the beginning of the sixteenth century, few universities had properly functioning medical faculties, the only important centre outside Italy being that at Montpellier. By the end of the eighteenth century, most European countries had at least one reputable medical faculty. This growth in university medical education was accompanied by an increasing awareness of the value of practical tuition: the first physic gardens were established at Pisa and Padua in the mid-sixteenth century; at the same time, bedside teaching was inaugurated at Padua; the first permanent anatomical theatres were set up at Padua and Leyden at the end of the sixteenth century. The expansion in practical medical education was particularly pronounced during the first decades of the eighteenth century; in this, the medical faculty at Leyden, with Herman Boerhaave (1668–1738) as its outstanding luminary, was pre-eminent. Lindeboom has summarized the situation as follows:

> ... before Boerhaave's appearance practical medicine was in a state of confusion and often not more than a precarious empiricism. The basic sciences of physics and chemistry were not yet applied to medicine in a well-balanced way, and they often gave rise to unfruitful speculation. Knowledge of anatomy and pathology was only poorly integrated into medicine; pathology received little attention and was without a comprehensive system. Clinical medicine was not well developed, and good bedside teaching was exceedingly rare. Moreover there was an obvious lack of reliable textbooks.[1]

Boerhaave's theoretical approach to medicine was based on the application of mechanics and hydrostatics; he belonged to the iatrophysical school, which had originated in Italy. He opposed the iatrochemical views introduced by Paracelsus (1492–1541) and promoted by Van Helmont (1577–1644), Franciscus Sylvius (1614–72), one of Boerhaave's predecessors at Leyden, and others. Nevertheless, Boerhaave maintained a lifelong interest in chemistry, contributing to its development as an independent science and ensuring for it a permanent place in medical education. In 1718 he was appointed Professor of Chemistry at the University of Leyden,[2] although he had already lectured on the subject privately for sixteen years. In this position, Boerhaave became head of the university laboratory, established in 1669, which he used for his public demonstrations and his chemical research. The accommodation in the laboratory had to be increased to handle the student numbers attracted by Boerhaave.

Boerhaave's influence on medical education, exerted through both his teaching and writing, was enormous. The medical school at Leyden attracted large numbers of students from other countries: during Boerhaave's thirty-year tenure as Professor of Medicine (1709–38), about a third of the 2000 students enrolled in the medical faculty were English-speaking.[3] Boerhaave's students carried his ideas to other European countries. His books were widely influential. Among these was a text on chemistry, which enjoyed great popularity from its first appearance, in an unauthorized edition, in 1724 until towards the end of the eighteenth century. In 1830 Thomson, Regius Professor of Chemistry at the University of Glasgow, assessed its significance as follows:

> [It] was undoubtedly the most learned and most luminous treatise on chemistry that the world had yet seen; it is nothing less than a complete collection of all the chemical facts and processes which were known in Boerhaave's time. . . . Every thing is stated in the plainest way, stripped of all mystery, and chemistry is shown as a science and an art of the first importance, not merely to medicine, but to mankind in general.[4]

Lindeboom and others have claimed that Boerhaave 'created the prototype of the present-day teaching programme and the medical curriculum'.[5] His medical course involved preliminary sciences, especially chemistry and botany, followed by classes in the structure and function of the human body, pathology and therapeutics, and, finally, clinical medicine. At the practical level, this involved the anatomical theatre, the physic garden, the bedside and the chemical laboratory.

The medical school at the University of Edinburgh, founded in 1726, was modelled on that at Leyden; all of the foundation professors had studied at Leyden. The outstanding success of the Edinburgh school is an important component of that historical phenomenon called the Scottish Enlightenment. The earliest teaching of chemistry at Edinburgh predates the foundation of the medical school but was, nevertheless, within a medical context.[6] Formal instruction, taking place in a laboratory, was provided by the Incorporation of Surgeon-Apothecaries at the beginning of the eighteenth century. In 1713 James Crawford (1682–1731) was appointed Professor of Physic and Chemistry at the University of Edinburgh. The few courses that he gave between 1713 and 1726 involved a large practical section in which several experiments, especially pharmaceutical processes, were demonstrated in a laboratory.[7] From the outset, chemistry was among the courses offered by the newly founded medical school. Initially, a number of professors were involved in the teaching of this subject, but from 1734 Andrew Plummer (1698–1756) assumed full responsibility. Instruction took place in a chemical laboratory set up in a house, which was owned by some of the medical professors, adjoining the University. Course content centred on the pharmaceutical application of chemistry, this providing a vocational training in drug preparation. Plummer's students watched him prepare drugs, which were then sold to apothecaries. As in the case of the other medical professors, Plummer's appointment by the Town Council[8] was for life, but was unsalaried. Income was derived from student fees and the sale of drugs;[9] from this, the professors equipped and maintained their laboratory. Plummer gave annual courses until 1755, when he was incapacitated by a stroke. At the end of that year, William Cullen (1710–90) was appointed Joint Professor of Medicine and Chemistry, becoming sole professor following Plummer's death in 1756.

1 Cullen, First Lecturer in Chemistry

In the autumn of 1744, Cullen moved to Glasgow from his native Hamilton, where he had been in practice. This city's university was part of its attraction; according to Cullen's biographer:

> To be the founder of a Medical School at Glasgow, similar to that which had been established at Edinburgh, was a design worthy of the active and enterprising genius of Dr Cullen, and he appears to have engaged in it with all the ability, diligence and ardour, necessary for the accomplishment of so important an object.[10]

Cullen began giving extramural medical lectures soon after his arrival in Glasgow, and, in the winter of 1746/7, by arrangement with Dr Johnstoun, the Professor of Medicine, he gave a course within the university on the theory and practice of physic.

At the beginning of the eighteenth century, the main function of the University of Glasgow, which had been founded in 1451, was the provision of clergy for the Presbyterian church. There were two chairs: a long-standing one in divinity and one in mathematics, dating from 1691. Teaching was carried out by regents, each of whom took entrants from one academic year through the entire four-year Arts course, consisting of Greek, logic, ethics and natural philosophy in successive years. The eighteenth century saw an impressive increase both in the range of subjects taught and in specialist teaching, a trend which began well before the final abolition of regenting in 1727: chairs were established in Greek (1704), divinity (second chair, 1705), humanity (1706), oriental languages (1709), civil law (1713), medicine (1713), ecclesiastical history (1716), anatomy (1720; for many years this was combined with botany), moral philosophy (1727), logic (1727), natural philosophy (1727), practical astronomy (1760); lectureships were established in botany (1704; in the same year the creation of a physic garden was proposed), chemistry (1747), materia medica (1766), midwifery (1790); at various times *ad hoc* arrangements were made for instruction in certain subjects.[11] While the beginnings of a medical school at Glasgow date from early in the eighteenth century, it did not begin to flourish until the mid-century:

> The country had reached a stage [early 1720s] at which a Medical School might have prospered, but its development was hindered by the aversion to teaching displayed by Johnstoun, the professor of Medicine, and Brisbane, the professor of Anatomy and Botany. Surgeons from the city were called in to teach Anatomy while Brisbane neglected that duty, and students seem to have been attracted in appreciable numbers; but it was not till Cullen and others had succeeded to the places of the early and indolent professors, and other needful subjects began to be taught, that the Medical School had a fair opportunity to develop.[12]

One of these 'needful subjects' was chemistry.

In 1744 Alexander Dunlop, son of the Professor of Greek and grandson of a former Principal of the University, was appointed Professor of Oriental Languages. At the time, he was in Geneva acting as tutor to the son of Sir James Campbell, and he delayed commencing his professorial duties until towards the end of 1745.[13] The salary which had accrued during his leave of absence was used to establish the teaching of chemistry within the University. The Minutes of the University Meeting of 5 January 1747 record the event:

> Mr Dunlop having represented the necessity of having Chemie taught in this University, he proposed at the same time that the thirty pounds sterling of the Professor of Oriental Languages salary that was saved during the time he was abroad with whatever the University should think fit to add thereto might be allotted to the buying the necessary apparatus for practical lectures in Chemie and building furnaces & c. for the same.[14]

Thereafter, events moved rapidly: an estimate of £52 for 'the Expenses necessary for an Apparatus for teaching Chemistry'[15] was submitted; a University Meeting, attended by Cullen and John Carrick (assistant to Robert Hamilton, Professor of Botany and Anatomy) in an advisory capacity, authorized the spending of this sum (£30 from the salary of the Professor of Oriental Languages and £22 from College revenues) under the direction of Cullen and Carrick;[16] a committee was set up to look into the conversion of the anatomy classroom into one for chemistry;[17] payment of £52 to Cullen and Carrick was arranged.[18] The first chemistry course was begun by Carrick in 1747, but, when he became ill after a few lectures, Cullen took over. Cullen continued to give chemistry courses during each College session until he moved to Edinburgh in 1756; he was succeeded by Joseph Black (1728–99), his pupil and friend.

The appointment of 'lecturers' at the University of Glasgow dates from the end of the seventeenth century, when William Jameson was asked to lecture on civil history. In 1704 John Marshall was appointed as keeper of the physic garden and teacher of botany. However, the title 'lecturer' does not appear systematically in the university records until 1766, when a 'Lecturer in Chemistry' was sought to replace Black.[19] Lecturers, unlike professors, did not have long-term tenure; their appointments were usually subject to annual renewal by the University Meeting or, from 1771, by the Senate or Faculty. Further, they did not share in university administration; unless they also held chairs,[20] they were not members of the University Meeting, Faculty or Senate. They did, however, have the same teaching responsibilities as professors and the same right to collect student fees.

In 1749 Cullen reported the successful introduction of chemistry teaching in the University; the University Meeting thanked him 'for the great care and pains he has been at in giving Chemical Lessons and explaining them constantly by the most usefull and necessary Chemical Processes and Experiments'.[21] The expense involved in the 1747 and 1748 courses exceeded the £52 allowed,[22] and the University Meeting acknowledged that items purchased by Cullen at his expense were his property. He reported his willingness to continue the course if the University would contribute to the expense involved;[23] a sum of £20 was allowed.[24] In 1751 Cullen reported that the expense of his chemistry course exceeded the income from student fees, and he requested that the allowance of £20 per session be continued for some years.[25] It was resolved to allow £20 for the next session, and thereafter to review the situation annually.[26] From 1751 to 1817 the lecturer in chemistry was annually granted an allowance: this was £20 until 1784, when Irvine's payment as Lecturer in Chemistry and Lecturer in Materia Medica was increased from £30 to £50;[27] from 1788 Hope received £50 as Lecturer in Chemistry;[28] from 1794 Cleghorn and later Thomson received £70,[29] this remaining unchanged until 1818, when Thomson became the first Regius Professor of Chemistry. As well as this annual allowance, the lecturers received student fees; in 1766, for example, Thomas Reid, Professor of Moral Philosophy, estimated Black's income from chemistry course fees to be £50 or £60.[30]

Cullen was critical of chemistry courses that were largely restricted to pharmaceutical considerations. His own courses were broadly based; he included discussion of the history, theory and industrial and agricultural applications of the subject.[31] Chemistry was presented as being significant not only in medicine, but also in eighteenth-century Scotland, and as being a subject worthy of study by polite society: Cullen's discussion of the causes of chemical change (his 'philosophical chemistry') was in harmony with the philosophical emphasis of the age; his discussion of the usefulness of chemistry[32] was in harmony with the improvement-based ideology of the age. In his teaching, Cullen stressed the practical aspects of chemistry. He used the £52 provided by the University and his own funds to equip a laboratory.[33] As was usual at the time, this was used in association with lecture demonstrations, which were an important component of Cullen's and other chemistry courses.[34] Less usual, and possibly unique, was the access to the laboratory which Cullen gave his students, thereby encouraging them to carry out their own practical work. The student response appears to have been less than enthusiastic; in concluding one of his courses, Cullen observed:

> . . . every one therefore should in the first place endeavour by easy experiments to acquire some knowledge in this way. The laboratory has been open to you but I am sorry to find that so few of you have frequented it. . . . Any of you that have a desire to go farther in the practice may have all the assistances the laboratory affords and you will perhaps find me at more leisure to assist you than I hitherto could.[35]

Outstanding among Cullen's chemistry students at Glasgow was Joseph Black, whose talents he quickly recognized; in an autobiographical memorandum, Black recalled:

> Dr Cullen began also at this time to give lectures on Chemistry, which had never before been taught in the University of Glasgow, and finding that I might be useful to him in that undertaking, he employed me as his assistant in the laboratory, and treated me with the same confidence and friendship and direction in my studies, as if I had been one of his own children. In this situation, I lived three years.[36]

Another student, Dobson, and the laboratory were involved in Cullen's celebrated discovery of cooling by evaporation; Cullen described the event as follows:

> . . . one of my pupils, whom I had employed to examine the heat or cold that might be produced by the solution of certain substances in spirit of wine, observed to me: That, when a thermometer had been immersed in spirit of wine . . . ; upon taking the thermometer out of the spirit, and suspending it in the air, the mercury in the thermometer . . . always sunk two or three degrees.[37]

Cullen was an excellent teacher with a paternal concern for the welfare of his students.[38]

In 1749 Johnstoun offered to resign as Professor of Medicine in favour of Cullen if he was permitted to retain his university house throughout his lifetime;[39] the University Meeting accepted this offer,[40] and, early in 1751,

Cullen was admitted to office.[41] Following his appointment as Joint Professor of Medicine and Chemistry by the Town Council (19 November 1755), Cullen began teaching in Edinburgh in January 1756, although he did not submit his resignation at Glasgow until the following March,[42] a delay that angered his patron, the 3rd Duke of Argyll.[43] In 1766 Cullen was appointed Professor of the Institutes of Medicine at Edinburgh (he was succeeded in the chemistry chair by Black), and in 1773 he became Professor of the Practice of Physic.

2 The Lecturers in Chemistry

The Lectureship in Chemistry was established, in fact but not in title, in 1747; the position lapsed with the appointment of the first Regius Professor of Chemistry in 1818. During this period there were seven lecturers in chemistry.[44] The most famous is Joseph Black, Cullen's pupil and successor, who was lecturer from 1756 to 1766.[45] Black's outstanding contributions to chemistry are his University of Edinburgh M.D. dissertation work on causticity, which is celebrated for the gravimetric approach employed and for its recognition of *fixed air* (carbon dioxide) as a gas chemically distinct from air, and his studies of heat, carried out at the University of Glasgow between 1756 and 1766. He left Glasgow to succeed Cullen as Professor of Medicine and Chemistry at Edinburgh, a position that he held until his death in 1799. All of the Lecturers in Chemistry after Black had attended his chemistry courses. The first of these was John Robison (1739–1805), the only lecturer not to have a medical degree, who held the position until 1769.[46] He later became Professor of Natural Philosophy at the University of Edinburgh (1774–1805), and, while there, he edited Black's *Lectures*.[47] William Irvine (1743–87) succeeded Robison; he retained the Lectureship in Materia Medica, to which he had been first appointed in 1766, although the University Meeting specified that 'the Union of the two Lectureships shall continue no longer than till qualified Persons can be found for each'.[48] The 'Union' did in fact continue until 1788. Irvine, who had been Black's student and research assistant, extended and modified Black's work on heat; his theory was widely influential at the end of the eighteenth century.[49] Following Irvine's death in 1787, Thomas Charles Hope (1766–1844) was appointed to both his lectureships;[50] the positions were separated in the following year, with Hope retaining the Lectureship in Chemistry, while Robert Cleghorn (1755?–1821) became Lecturer in Materia Medica.[51] In 1789 Hope was appointed 'Assistant & Successor'[52] to his uncle, Alexander Stevenson, Professor of Medicine at the University of Glasgow, but he continued as Lecturer in Chemistry until Stevenson's death in 1791, when he became Professor of Medicine.[53] While at Glasgow, Hope's outstanding contribution to chemistry was his analysis of a mineral found near the village of Strontian in Argyllshire; he found that it contained a peculiar 'earth', which he called *strontia*. (The element itself, strontium, was first isolated by Davy in 1808.) In 1795 Hope moved to

Edinburgh to become Conjoint (with Black) Professor of Medicine and Chemistry; he became sole Professor following Black's death in 1799, a position that he held until 1843. As Lecturer in Chemistry, Hope was succeeded by Cleghorn, who held the position for 26 years.[54] He was an eminent physician who, according to Emerson,[55] failed to secure a Glasgow medical chair in the 1790s because of the opposition of the Dundas political machine. The last Lecturer in Chemistry and first Regius Professor of Chemistry was Thomas Thomson (1773–1852),[56] whose contributions to chemistry were many: about 200 publications covering many aspects of the subject; a famous textbook, *A System of Chemistry*, first published in 1802; a work on the history of chemistry; his role as Dalton's bulldog; his teaching of the subject, particularly its practical aspects. Thomson continued as Regius Professor until his death in 1852, but, during the last decade of his tenure, he was increasingly assisted by his nephew, Robert Dundas Thomson.

3 The Chemical Laboratory

Accommodation

Initially, the laboratory was housed in an existing room within the University. In 1763, in justifying the building of a new laboratory, Black informed the University Meeting:

> ... that the present Laboratory is by no means sufficient for all necessary & usefull purposes, that it is not sufficiently large for carrying on all the proper experiments, & is so damp & disagreeable in its present condition as would deter many students from attending the Lectures on Chemistry were they to be delivered there, the floor never having been laid nor the walls plaistered in any manner & that the Teacher of Chemistry has been therefore subjected to the necessity of Teaching in another room while the Processes are going on in the Laboratory.[57]

Because of a problem in accommodating the mathematics class, it was resolved, in 1763, to build a new laboratory, at an expense not exceeding £350, and to convert the existing laboratory into a mathematics classroom.[58] There were financial and constitutional objections to building 'a spacious & ornamental Laboratory' for a subject that 'has no Foundation in the College'.[59] Nevertheless, a new laboratory was built at a cost of almost £368;[60] it appears to have been completed just prior to Black's departure from Glasgow. During the following decades, the University Meeting, or, from 1771, the Faculty, approved several grants to maintain and to alter the laboratory. (In 1801 Cleghorn incurred expense without Faculty permission, a lapse in judgement with which many of us today would have considerable sympathy. He received payment following written and verbal apologies.[61]) In

Faculty minutes from the beginning of the nineteenth century, the laboratory
is usually referred to as the chemistry class or the chemistry class room; it was
the place where Cleghorn gave lectures illustrated by lecture demonstrations.
This was moved to a new location in 1810/11.[62] Thomson was later to claim
that: 'There was no laboratory when I came here [1817] at all.'[63] Within
months of his arrival, Thomson submitted recommendations which resulted
in the existing chemistry classroom being converted into a laboratory, this
including a partitioning to provide an apparatus room, and the logic classroom being converted into a teaching room for chemistry.[64] With Thomson
the function of the laboratory underwent an important change.

Equipment

Detailed information about equipment (and mineral and chemical collec-
tions) in the laboratory at various times can be obtained from accounts and
inventories.[65] The University would occasionally allow funds for equipment
maintenance and purchase; in 1805, for example, Cleghorn was granted a sum
not exceeding £30 for a galvanic apparatus.[66] It was usual for the lecturers to
provide apparatus from their own funds, and, in some cases, the University
purchased some or all of this at the completion of a lecturer's tenure.[67]

Lecture Demonstration

During the eighteenth century, Scotland enjoyed that intellectual efflores-
cence now referred to as the Scottish Enlightenment. An important aspect of
this was the excellence of much of the university teaching, a situation
undoubtedly encouraged by a system in which academic incomes were
derived largely or entirely from student fees. The Lecturers in Chemistry
appear to have kept their courses up to date, and their course content was
sometimes outstandingly innovative: Cullen, Black and Irvine incorporated
the results of their research work in their courses; Hope began teaching
Lavoisier's 'new system of chemistry' as early as 1788;[68] Thomson was
probably the first to teach Dalton's ideas. The standard method of teaching
chemistry throughout the eighteenth century was the lecture accompanied by
lecture demonstrations. This was the approach used at Glasgow during the
period of the lectureship, the course usually taking place in the laboratory,
where the lecturer would carry out many chemical processes illustrating
points made during the lectures. In two sketches of Black lecturing,[69] we can
see him standing before a large table, on top of which there are his notes and
apparatus, under the gaze of his students. A well-known verbal image was
provided by Lord Brougham, who attended Black's 1796 course at Edinburgh:

Nothing could be more suited to the occasion; it was perfect philosophical
calmness; there was no effort; it was an easy and a graceful conversation. The

voice was low, but perfectly distinct and audible through the whole of a large hall crowded in every part with mutely attentive listeners. . . . In one department of his lecture he exceeded any I have ever known, the neatness and unvarying success with which all the manipulations of his experiments were performed. His correct eye and steady hand contributing to the one; his admirable precautions, foreseeing and providing for every emergency, secured the other. I have seen him pour boiling water or boiling acid from a vessel that had no spout into a tube, holding it at such a distance as made the stream's diameter small, and so vertical that not a drop was spilt. . . . The long table on which the different processes had been carried on was as clean at the end of the lecture as it had been before the apparatus was planted on it. Not a drop of liquid, not a grain of dust remained.[70]

We can imagine Black carrying out experiments as he lectured:

I pour on the Iron filings some of the Vitriolic Acid, diluted w.^{t} an equal q.^{ty} of Water, yet it will not dissolve quickly, though this q.^{ty} of water is added, therefore, we shall add more water in order to bring on the Action of the Acid.- W_^{n} a due degree of Solution is obtained, we shall find the Acid act quickly & the Mixture will be heated.-
You now Observe it goes on w.^{t} a considerable Effervescence w.^{c} are disagreeable to the Smell.- . . . on applying a lighted Candle to it you see the Vapour immediately takes fire. . . . W_^{n} this dissolution is finished it proves a liquor of a pale green Colour.[71]

In such a manner Black and other lecturers described and demonstrated hundreds of chemical processes during their courses. (Swinbank[72] has discussed the teaching of other experimental sciences at the University of Glasgow in the eighteenth century.)

4 Practical Classes

The end of the eighteenth century and the early nineteenth century saw the appearance and development of laboratory-based practical classes and the associated growth of research schools. This radical change in the teaching of chemistry was an important factor in the nineteenth-century professionalization of science. Pre-eminent in this transformation was Liebig's chemical laboratory at Giessen, which was established in 1824. In Britain the leader was Thomas Thomson.

At the Universities of Edinburgh and Glasgow early in the nineteenth century, Hope and Cleghorn maintained the lecture demonstration tradition. In this, Hope, like his predecessor Black, was a master, attracting enormous student numbers and enjoying wide acclaim. But, as one of his students later pointed out, 'Dr Hope, with all his ability as a teacher, made very few chemists, because he never encouraged practical study';[73] 'His laboratory was open to no one but his class assistant'.[74] At Edinburgh extramural lecturers offered practical classes, the first to do so being Thomson from 1807 to 1811,[75] and students formed chemical societies in which they performed experiments.[76] It was not until 1823 that practical classes were offered in the University of Edinburgh.[77] In a letter written in 1817, the year of his appointment as Lecturer in Chemistry at Glasgow, Thomson stated his ambition, 'to erect a laboratory upon a proper scale to establish a real chemical school in Glasgow and to breed up a set of young practical chemists'.[78] With the availability of the new laboratory in 1818, Thomson began a practical class. This was limited to up to ten students,[79] who worked in the laboratory between three and four hours a day, six days a week during the ten-month course.[80] Thomson described his method of practical instruction as follows:

> I have a set of rules, which are in manuscript, in the laboratory; [the students] make themselves quite masters of those rules, and I give them different minerals to analyse; they continue in the laboratory . . . till they become expert chemists, and there are five or six that have gone out of my laboratory that I think are as good chemists as any men in existence.[81]

Thomson's practical class was the first in a British university.

Thomson was also ambitious to create a research school in his teaching laboratory. To quote Morrell:

> It appears then that between 1819 and 1835 Thomson ran a research school focused on three related aims: to put Dalton's atomic theory on a wider and firmer experimental basis; to provide conclusive experimental evidence to substantiate Prout's hypothesis that the atomic weights of elements were whole-number multiples of that of hydrogen; and to determine the chemical composition and formulae of all known minerals, particularly those containing aluminium.[82]

For a variety of reasons, analysed by Morrell,[83] the Thomson research school was very much less successful than that of Liebig, which was centred on organic chemistry. Nevertheless, in introducing practical classes and launching a research school at the University of Glasgow, Thomson led the way in Britain. The growing number of students attending his chemistry courses necessitated the erection of a new chemistry building, at a cost of about £5000. This was located in Shuttle Street, outside the College walls, and was opened in 1831. The chemistry department, occupying the three top floors, included a large teaching laboratory, which enabled Thomson to expand his practical classes and to develop the teaching of qualitative and quantitative analysis.[84] In 1852 Thomson reported that 'The Chemical Class room & laboratory are both excellent. But the whiskey shop which occupies the ground floor does not accord with what one would expect from the University of Glasgow.'[85] The Shuttle Street building continued to house the chemistry department until 1870, when the University of Glasgow was moved to its present location on Gilmorehill.

5 Conclusion

Taking the University of Glasgow as a case study, the role of the laboratory in university chemistry has been discussed. From the outset, the laboratory has been linked to the teaching of chemistry, although its function has changed substantially. This can be seen as part of an increasingly practical emphasis in medical education, although the introduction of chemistry teaching in universities was by no means always in a medical school context. Until the early nineteenth century, the teaching function of the laboratory was related to the lecture demonstration. Few seem to have been as enlightened as Cullen in encouraging students to carry out their own practical work. The laboratory was also used for the teacher's research and consulting activities. The familiar pattern of practical classes and research schools emerged early in the nineteenth century.

Acknowledgements

This chapter is part of a study of chemistry during the Scottish Enlightenment completed while I was an Honorary Senior Research Fellow in the Department of Chemistry, University of Glasgow, and an IPSE Fellow in the Institute for Advanced Studies in the Humanities, University of Edinburgh. I gratefully acknowledge the award of these Fellowships and the assistance of Professor S. J. Thomson, Mr M. Moss and Dr D. Dow.

Notes and References

The following abbreviations have been used: *GUA*: Glasgow University Archives. *MUM*: Minutes of University Meetings (University of Glasgow). *MMF*: Minutes of Meetings of Faculty (University of Glasgow). *MRUM*: Minutes of Rector's Meetings and University Meetings (University of Glasgow). *MMS*: Minutes of Meetings of Senate (University of Glasgow).

1. Lindeboom (1968), p. 9.
2. Boerhaave's appointments at the University of Leyden: 1701, Lecturer in Medicine; 1709, Professor of Botany and Medicine; 1718, Professor of Chemistry. From 1718 to 1729, when he resigned as Professor of Botany and Chemistry, Boerhaave held three of the five chairs that constituted the faculty of medicine.
3. Comrie (1939), p. 32. During this period 1919 students were enrolled; of these, 659 were English-speaking (340 from England, 205 from Scotland, 107 from Ireland and 7 from the British colonies). Boerhaave's influence in Great Britain is discussed in Lindeboom (1974).
4. Thomson (1830), Vol. 1, p. 212.
5. Lindeboom (1968), pp. 360–1.
6. Anderson (1978), p. 3.
7. *Ibid.*, pp. 4–5.

8. The University of Edinburgh was controlled by the Town Council from its foundation in 1583 until the 1858 Universities (Scotland) Act.
9. Anderson (1978), pp. 8–9.
10. Thomson (1832–41), Vol. 1, p. 24.
11. Coutts (1909).
12. *Ibid.*, p. 186.
13. *Ibid.*, p. 239.
14. *MUM* 1730–49 (GUA 26639) (Meeting 5 January 1747), pp. 224–5.
15. *Ibid.* (Meeting 28 January 1747), pp. 226–7. This estimate is in Glasgow University Archives, GUA 5409.
16. *Ibid.* (Meeting 11 February 1747), p. 229.
17. *Ibid.* (Meeting 1 April 1747), p. 230.
18. *MMF* 1745–53 (GUA 26649) (Meetings 18 June 1747, 25 June 1747, 3 February 1748), pp. 31, 35, 49.
19. *MUM* 1763–8 (GUA 26643) (Meeting 9 June 1766), p. 131.
20. Cullen was appointed Professor of Medicine in 1751. Black was appointed Professor of Botany and Anatomy in 1756; he resigned from this chair in 1757 and was then appointed Professor of Medicine. Hope vacated the lectureship in chemistry on being appointed Professor of Medicine in 1791.
21. *MUM* 1730–49 (GUA 26639) (Meeting 26 June 1749), p. 11.
22. The total sum expended was £136. 7s 10d. A detailed account of this expenditure is in Glasgow University Archives, GUA 5409.
23. *MUM* 1730–49 (GUA 26639) (Meeting 1 November 1749), pp. 14–15.
24. *Ibid.* (Meeting 14 November 1749), p. 17.
25. *Ibid.* (Meetings 20 February 1751, 27 March 1751, 5 April 1751), pp. 52–4. A letter sent by Cullen concerning this matter is in Glasgow University Archives, GUA 31115.
26. *Ibid.* (Meeting 29 May 1751), p. 63.
27. *MMF* 1780–4 (GUA 26692) (Meeting 10 June 1784), p. 8.
28. *MMF* 1784–9 (GUA 26693) (Meeting 10 June 1788), p. 313.
29. *MMF* 1789–94 (GUA 26694) (Meeting 10 June 1794), p. 367.
30. T. Reid, letter to Dr David Skene (Glasgow, 18 April 1766). In Hamilton (1863), Vol. 1, pp. 45–6.
31. Anderson (1978), p. 11.
32. Cullen was by no means the first to emphasize the usefulness of chemistry. In England, prior to the eighteenth century, Bacon, Boyle (and other early Fellows of the Royal Society) and others had indicated the practical importance of science. In the 1720s and 1730s, Peter Shaw (in his abridged editions of the works of Bacon and Boyle, his English editions of Boerhaave and Stahl, his essays and his lectures) highlighted the usefulness of chemistry (Gibbs, 1951b, 1952b). He was followed by William Lewis (Gibbs, 1952a) and Robert Dossie (Gibbs, 1951a). Cullen referred to the work of Shaw, Lewis and Dossie. Chemistry courses given at Cambridge in the 1750s and 1760s by John Hadley and Richard Watson, successive Professors of Chemistry, show an appreciation of the practical significance of the subject (Coleby 1952, 1953).
33. *MUM* 1730–49 (GUA 26639) (Meeting 26 June 1749), p. 11.
34. In the mid-eighteenth century in British universities, lecture demonstrations were used by Plummer at Edinburgh and by Hadley and Watson at Cambridge.
35. Quoted in Anderson (1978), p. 12.
36. Quoted in Ramsay (1918), p. 4.
37. Cullen (1756), p. 145.
38. *The Bee, or Literary Weekly Intelligencer* (1791), **I**, 1–10, 45–56, 121–5, 161–6. This series of articles, *Cursory Hints and Anecdotes of the late Doctor William Cullen of Edinburgh*, published the year after Cullen's death by a former

chemistry student and an acquaintance of more than 30 years, lavish praise on Cullen's abilities as a teacher and concern for his students.

39. Letter from Johnstoun to the Rector, 30 October 1749, GUA 30297.
40. *MUM* 1749–59 (GUA 26640) (Meeting 1 November 1749), pp. 14–15.
41. *Ibid.* (Meeting 2 January 1751), p. 47.
42. *Ibid.* (Meeting 22 March 1756), pp. 170–1. In his resignation (dated 22 March 1756), Cullen stated that 'the said office was really and freely by me vacated renounced and resigned from and after the twelfth of January last . . .'.
43. At Glasgow and in his first appointment at Edinburgh, Cullen enjoyed the patronage of the 3rd Duke of Argyll, political manager for Scotland from 1746 until his death in 1761. (Patronage in eighteenth century Scotland has been discussed by Roger L. Emerson in 'The structure of university patronage in Scotland, 1690–1780', a paper presented at the British Society for the History of Science Meeting on *The Patronage of Science*, St. Hilda's College, Oxford, 14–16 July 1986.) Cullen's procrastination in resigning his Glasgow chair, following his appointment at Edinburgh, angered the Duke of Argyll; this is discussed in a letter sent by William Ruat to Robert Simson (London, 11 March 1756), GUA 30492. The delay may have been associated with Plummer's initial refusal to co-operate with Cullen because he had not been consulted concerning the latter's appointment; a compromise was reached in March, 1756 (Anderson, 1978, p. 13).
44. Kent (1950).
45. Black was appointed Professor of Botany and Anatomy on 22 March 1756 (*MUM* 1749–59 (GUA 26640) (Meeting 22 March 1756), pp. 170–3); he resigned this position to be appointed Professor of Medicine the following year (*Ibid.* (Meeting 8 April 1757), pp. 218–19). He lectured on chemistry throughout his ten-year tenure at Glasgow. He resigned as Professor of Medicine in 1766 (*MUM* 1763–8 (GUA 26643) (Meeting 27 May 1766), p. 120).
46. Robison was appointed Lecturer in Chemistry, on Black's recommendation, on 11 June 1766; at the same University Meeting, Irvine was appointed the first Lecturer in Materia Medica (*MUM* 1763–8 (GUA 26643) (Meeting 11 June 1766), pp. 141–2). Robison's resignation was received at a University Meeting on 16 May 1769; at the same Meeting he was rebuked for a breach of the peace and fined two guineas following an incident in which he struck a student, David Woodburn, with his stick (*MRUM* 1768–70 (GUA 26644) (Meeting 16 May 1769), pp. 89–91).
47. Black (1803).
48. *MRUM* 1768–70 (GUA 26644) (Meeting 11 June 1770), pp. 191–2.
49. Fenby (1987).
50. *MMF* 1784–9 (GUA 26693) (Meeting 10 October 1787), p. 257.
51. *Ibid.* (Meetings 16 May 1788, 10 June 1787), pp. 305, 306, 313.
52. *MMF* 1789–94 (GUA 26694) (Meeting 23 October 1789), pp. 21–2.
53. *Ibid.* (Meetings 3 June 1791, 10 June 1791), pp. 157, 165.
54. Cleghorn's appointment as Lecturer in Chemistry: *ibid.* (Meeting 10 June 1791), p. 165. His resignation: *MMF* 1813–25 (GUA 26698) (Meeting 8 August 1817), p. 135.
55. Emerson, *op. cit.* [Note 43].
56. Thomson's appointment as Lecturer in Chemistry: *MMF* 1813–25 (GUA 26698) (Meeting 5 September 1817), p. 136. His appointment as Regius Professor of Chemistry: *MMS* 1802–19 (GUA 26688) (Meetings 12 March 1818, 17 March 1818), pp. 340–2.
57. *MUM* 1763–8 (GUA 26643) (Meeting 31 October 1763), p. 8.
58. *MUM* 1760–3 (GUA 26642) (Meetings 20 April 1763, 14 May 1763), pp. 229, 230, 234; *MUM* 1763–8 (GUA 26643) (Meetings 24 June 1763, 28 October 1763, 31 October 1763, 2 November 1763, 22 May 1764), pp. 2, 5–9, 28.
59. MUM 1763–8 (GUA 26643) (Meeting 28 October 1763), pp. 5–7.

60. University records show three payments to the builder, Mr Michael Bogle: *MMF* 1753–5 and 1761–70 (GUA 26650) (Meetings 29 October 1764, 14 June 1765), pp. 75, 94; *MUM* 1763–8 (GUA 26643) (Meeting 10 June 1766), pp. 132, 133. (For comparative purposes, it is perhaps relevant to point out that an observatory built in the late 1750s cost about £400: Coutts (1909), p. 229.)

61. *MMF* 1800–6 (GUA 26696) (Meetings 11 May 1801, 10 November 1801, 19 November 1801), pp. 41, 69, 70.

62. *MMF* 1806–13 (GUA 26697) (Meetings 27 March 1810, 3 April 1810, 3 May 1810, 10 October 1810, 19 December 1810, 11 June 1811), pp. 292–4, 296, 298, 320, 331, 358, 359.

63. *Evidence, Oral and Documentary, taken and received by the Commission appointed by His Majesty George IV, July 23d, 1826; and re-appointed by His Majesty William IV, October 12th, 1830; for visiting the Universities of Scotland,* vol. II, *University of Glasgow,* London, 1837, p. 203.

64. *MMF* 1813–25 (GUA 26698) (Meetings 12 March 1818, 26 March 1818, 5 May 1818, 30 October 1818), pp. 148, 149, 154, 159, 160.

65. a. GUA 5409 An account of Cullen's expenses in giving the 1747 and 1748 chemistry courses.
    b. GUA 43081 Inventory of equipment received by Robison on becoming Lecturer (dated 28 October 1766); this list includes a list of equipment purchased from Black (*MUM* 1763–8 (GUA 26643) (Meeting 28 October 1766), p. 168).
    c. GUA 43082 Inventory of equipment received by Irvine (dated 13 July 1769).
    d. GUA 5190 Inventory of equipment left by Cleghorn (dated 21 January 1818).

66. *MMF 1800–6* (GUA 26696) (Meetings 7 March 1805, 31 May 1805, 10 June 1805, 17 January 1806), pp. 351, 359, 362, 381. Also GUA 5179.

67. Apparatus left by Black was purchased for £12 9s 0d (*MUM* 1763–8 (GUA 26643) (Meetings 12 June 1766, 28 October 1766), pp. 146, 168). Black also took apparatus with him to Edinburgh and continued to claim apparatus after he had left Glasgow (Anderson, 1978, pp. 21, 22). Irvine's apparatus and mineral and chemical collection was purchased for £100 (*MMF* 1784–9 (GUA 26693) (Meetings 10 December 1787, 10 June 1788), pp. 268, 269, 339; *MMF* 1789–1794 (GUA 26694) (Meeting 1 November 1790), p. 90). It is interesting that the Faculty later granted an annuity of £50 to Irvine's widow (*MMF* 1813–25 (GUA 26698) (Meeting 13 January 1825), pp. 403, 404). Cleghorn's apparatus was purchased for £60 (GUA 5189, 5190; *MMF* 1813–25 (GUA 26698) (Meetings 22 October 1817, 18 February 1818), pp. 138, 144).

68. Donovan (1979).

69. Cochrane (1966), pp. xvi, xxviii, which reproduce these sketches.

70. Brougham (1855), p. 20.

71. Cochrane (1966), p. 140.

72. Swinbank (1982).

73. Christison (1885–6), Vol. 1, p. 62.

74. *Ibid.*, p. 58.

75. Morrell (1969b).

76. Christison (1885–6), Vol. 1, pp. 58–60.

77. *Ibid.*, p. 58.

78. Quoted in Morrell (1972), p. 2.

79. *Evidence, Oral and Documentary,* 1837, *op. cit.* [Note 63], Vol. II, p. 203.

80. *Ibid.*, p. 151.

81. *Ibid.*, p. 203.

82. Morrell (1972), p. 11.

83. *Ibid.*

84. Morrell (1969a).

85. Thomson, letter (Glasgow College, 24 January 1852) (GUA 3414).

3

Amusement Chests and Portable Laboratories: Practical Alternatives to the Regular Laboratory

Brian Gee

1 The Chemical Amusement Chest

One of the many tasks carried out by Sir Francis Galton — that tireless Victorian geographer, meteorologist, anthropologist, psychologist, statistician, biometrician and general jack of all science — was to inquire of some 180 eminent persons how they came to develop scientific interests in their youth.[1] Several replied by relating how their first contact had arisen through chemical chests and portable laboratories. Taking this clue beyond the anonymity of Galton's respondents, further insight may be gained from the obituary notices of Fellows of the Royal Society. Two examples alone are considered here.

The obituarist of the Sheffield University chemist William P. Wynne, in recording his school life in the 1870s, wrote:

> There was no school laboratory in which the boys could work, but Wynne improvised experiments at home with a 'Statham's Chemical Cabinet' and became fired with ambition to become a chemist.[2]

Similarly, the obituarist of the Leeds University chemist Arthur Smithells recalled:

> . . . it was apparently the casual purchase of *Statham's Chemical Cabinet* that first directed his thoughts to chemistry. The magic that could be produced from this shilling box of chemicals and test-tubes made a never-to-be-forgotten impression on his boyish mind. The lure of this spectacular introduction to chemistry did not fade with time, but was developed by reading a few books of science lectures.[3]

It is true that the boys in Wynne's class at the King Edward school in

Birmingham had been shown what could be done with a few pieces of simple apparatus by their teacher George Gore, himself a self-taught chemist who later became an FRS, and that Smithells likewise derived enthusiasm from the reading of H. E. Armstrong's Manchester lectures. But also it is a salutary reminder that school teaching laboratories were by no means the norm at this time.

The passages quoted above are particularly useful in identifying Statham's amusement chests as one source of influence, although it is perhaps dangerous to infer from such isolated examples that the chemical amusement chest held any importance beyond a mere juvenile introduction to the subject.[4] And yet, a prefatory remark in W. E. Statham & Sons' *Second Steps in Chemistry* reports the sale of over one-quarter million copies of *First Steps in Chemistry* (the tract which accompanied each chemical chest) in the previous century.[5] This amazing number offers some indication of the extent to which popular chemistry had pervaded Victorian culture. As to the origin of the juvenile amusement chest, an extant Statham trade catalogue published around the time of the 1862 London International Exhibition, with the proprietor himself writing in the third person, reveals:

> Mr Statham has reason to know that they [the amusement chests] have not only afforded a source of amusement and instruction for nearly a quarter of a century, but that much of the taste for chemistry which has prevailed for many years past, may be traced to their silent influence on the rising generation.[6]

From this it may be deduced that the market in juvenile amusement chests had taken off some time during the 1830s. More precisely, it had begun somewhat earlier when Frederick Accum, the Soho supply house chemist, delivered his popular lectures to a fashionable audience at the Surrey Institution in 1817.[7] 'So singular and curious are the phenomena of science', writes Accum, 'that to be welcomed and cultivated, it needs only be known.' Thus enticed, individuals could purchase an Accum amusement chest for prices between ten and eighteen guineas and try the spectacle of chemistry for themselves. Of this accompaniment, the manufacturer wrote:

> The object of this chest is to blend chemical science with rational amusement. The pleasing appearances which chemistry affords, whilst they amuse, by presenting to the observer striking phenomena and unexpected results, are well calculated to diffuse mirth and surprise through a friendly circle.[8]

Encouraging the 'Young Chemist to contemplate phaenomena [beyond] the mere amusement of a leisure hour' was the intention behind Accum's accompanying compendium of *Chemical Amusements*.[9]

Cheaper versions of amusement chest became available around the time of Faraday's first Christmas lectures, over 1827/8, when 'philosophical juveniles' were treated to 'tangible chemistry' by a performer *par excellence*.[10] For prices between 15s. and 30s., either at John Newman's in Regent Street or at

William Cary's in the Strand, the youth of London could then buy simple chests, although without any directive literature.[11] This deficiency was made good by another instrument maker, the optician Francis West of Fleet Street, who in 1830 issued boxes of philosophical and chemical goods together with accompanying literature described under the guise of 'intellectual toys'. According to one reviewer, West's tracts were 'replete with amusing information, conveyed in an agreeable manner, and adapted to juvenile minds'.[12] At last, for thirty shillings plus the sixpenny tract of *Chemical Recreations*, a London-bound youth could savour the subject in an orderly fashion with safety, so the parents were assured.[13]

Parents of this particular 'new age' not only had a long historical backdrop in the popularisation of science, but also reacted positively to Henry Brougham's cry for a scientific approach in the daily lives of ordinary folk, a philosophy that had been promulgated by the Society for the Diffusion of Useful Knowledge.[14] Moreover, they had been pounded by three decades of 'right-minded' Edgeworthism in which, in educational terms, it was argued that children needed practical things to do.[15] Thus, the commencement of juvenile lectures at the Royal Institution and West's opportunism was a reflection of these times. West himself declared his intellectual toys to be 'according to the recommendation of Miss Edgeworth' and, in a similar vein, another reviewer, in the *Mercantile Journal*, echoed the gist of the opening lines from the Edgeworths' famous *Practical Education*, by reference to the 'useless baubles and expensive toys' of times past.[16] In the opinion of the Edgeworths, children of the rising industrial age ought not be spoilt with wasteful *representational* toys of the kind offering a Lilliputian version of the world of adults (traditionally the dolls' house for girls and coach-and-six for boys), but rather their attention was to be engaged with *practical* toys from 'rational toyshops'. Neither the Edgeworths nor Thomas Beddoes before them, however, had much success in replacing ordinary toyshops with this form of outlet, although the concept in 1830 was sufficiently strong for the instrument trade to make such rational provisions for the juvenile market.

From the 1830s, then, the retailers of science, represented by West in particular, but later others such as W. E. Statham, R. B. Ede and Edward Palmer, were influential in spreading science among the young. 'The cultivation of the youthful mind in the various branches of science, is now happily rendered an object of solicitude and become a part of education', West told his customers, urging them onwards to 'lead the minds of their children towards the goals of science'.[17] That message rang loud and clear over the decades and by the 1860s the supplier Messrs Joseph, Myers, & Company could still influence parents of Elementary School children by informing them that 'A considerable amount of knowledge may be conveyed to the young by the medium of Scientific Toys and Models'. Likewise, Statham, by this time with more than twenty-five years of trade experience, could tell how his chemical amusement chests would lay 'in a pleasing and amusing manner, the foundations of the more dry and technical instruction of the schoolroom, class, or college'.[18]

It is in this setting that the portable laboratory, vulgarized in the form of a chemical amusement chest, offered youngsters such as Wynne and Smithells their first contact with chemical science and its necessary manipulations. Later, towards the end of the century, the chemical amusement chests competed with other technological arrangements such as electroplating sets, air-pump kits, electric machines, and model steam engines and the host of other (often imported) 'rational' or 'intellectual' toys which could be found in shops such as the Economic Electric Supply Company in the Edgware Road or Parkins and Gotto's in Oxford Street or, of course, from Statham's emporium at the 'Depôt for Scientific Amusements, Presents, &c' in the Strand.[19] But the amusement chest remained supreme in the mind of many children who, perhaps like the youthful Faraday, imagined that one day they might become scientists and work in a 'real' laboratory.

2 Early Portable Laboratories

Portable laboratories were by no means a new idea in the 1830s, when they made a wide-scale appearance in the form of amusement chests. Their appearance then may now be regarded as an inevitable — if not entirely satisfactory — response to a curriculum need at a time when teaching laboratories were, by and large, still a thing of the future. The earliest example appears to be that in Peter Shaw's *Three Essays in Artificial Philosophy or Universal Chemistry* (1730), in which the author states that a 'Portable Laboratory, ready fitted for Business, may be had of Mr Hauksbee, in Crane Court, Fleet Street, to facilitate and promote the Practice of Chemistry by putting a commodious Laboratory into the hands of Gentlemen'.[20] This arrangement, usefully accompanied by an *Essay for Introducing a Portable Laboratory* (1731), was aimed at those men of arts and manufactures following the Shaw–Hauksbee Jr. lecture-demonstrations.

The concept of portability itself was better appreciated later in the century by those ubiquitous travellers, the mineralogists, who found that certain operations could be concluded more speedily if performed in the field rather than at home, where chemical cabinets were normally housed. Boxes for carrying blowpipes and test reagents were used by A. F. Crönstedt in Sweden about 1770. Similarly, Guyton de Morveau in France described a *nécessaire chymique* in 1783, comprising two boxes, each measuring 7 in × 4 in × 1.5 in, one containing chemicals and the other containing hardware, including a blowpipe, tweezers, magnetizing needle, etc. Another more extensive arrangement appeared in Jena and was designed by J. F. A. Göttling in 1788 for use by analytical chemists, natural philosophers, mineralogists, metallurgists and those wishing to test mineral waters or wines for adulteration. For three golden louis a purchaser would have proudly owned two boxes, each

12 in × 9 in × 9 in, containing thirty-five chemicals, scales and weights, a blowpipe and miscellaneous glass apparatus.[21]

From these Continental developments the concept took hold in England. A translation of Crönstedt's work by Magellan in 1770 indicates that the same two boxes described by the former could be purchased from the General Office of Business, Arts and Trade, at 98 Wood Street in Cheapside; correspondingly, a translation of Göttling's book by the publishers C. & G. Kearsley tells of portable laboratories obtainable from their premises at 46 Fleet Street.[22] Customers for these arrangements formed the nucleus of British chemists and mineralogists, whose interests were very much tempered by the technical developments springing from the Paris School of Mines. Included were men such as William Babbington, who was particularly concerned with the classification of minerals; Bryan Higgins, founder of the Society of Philosophical Experiments and Conversations (founded 1794); and later members of the Askesian Society (founded 1796), the Royal Institution (founded 1799) and the Mineralogical Society (founded 1800).[23]

No less interested in the vogue of chemical mineralogy, although working at some considerable distance from the London-based fraternity, was William Henry at Manchester. He appears to have been the first (after some discontinuity with the Shaw–Hauksbee initiative) to have organized cabinets for learning from his magnesia works. In his *Epitome of Chemistry* he reveals:

> An object, which I propose to be fulfilled by this epitome, is, that it may serve as a companion to the collections of chemical substances, which I have been induced, by the repeated application of students of this science, to fit up for public sale.[24]

The second and third editions of this text contain advertisements detailing portable laboratories, selling at fifteen, eleven and six and a half guineas, which he claimed to be 'invented by' himself. See Table 3.1.

One person known to have purchased a fifteen guinea chest was James Watt junior, to whom Henry apologizes for its production:

> The bottles, I am sorry to confess, are not what I could wish; and indeed I had so much trouble in getting them at all (as well as the chests) in the country, that unless I can make an arrangement for procuring both from London, I shall not fit up any more.[25]

Indeed, in 1806 Henry reported his intention to hand over this sideline to Richard Knight of Foster Lane (London), a fellow member of the Mineralogical Society and seller of iron laboratory furnaces.[26]

Few descriptions of individuals using their portable laboratories are known. It is, therefore, worth recording the instance of Humphry Davy, who certainly was not to be without the means of experimenting during his travels through France and Italy in 1814. In his *Consolations on Travel* he records that 'all the wants of the philosophical chemist could be carried in a small

Table 3.1   Advertisement from Henry (1806).

DESCRIPTION AND PRICES

of the

PORTABLE CHEMICAL CHESTS

invented by William Henry,

and sold by him at his Laboratory in Manchester

No. 1.   A large double mahogany chest, with folding doors, containing eighty-six strong square bottles, with ground and cut stoppers, filled with tests, &c. and so arranged that the labels may be seen at one view; together with five drawers, in which are various articles of apparatus; accurate scales, decimal weights, improved blow-pipe and spoon .. .. ..

15 guineas.

No. 2.   A chest of similar construction, but smaller; containing fifty-two bottles, with other articles as in No. 1

11 guineas

No. 3.   An upright chest, box shaped, intended chiefly as a travelling companion; holding thirty-six bottles, with a drawer, containing articles and apparatus, as in the two foregoing ones.

6 guineas and a half.

N.B.   Any one of the foregoing chests may be had either with or without the scales and blow-pipe. If sold without scales and weights, 16 shillings may be deducted from the foregoing prices; and 14 for the blow-pipe and spoon.

The tests are all prepared with the most scrupulous attention to accuracy; and supplies may be obtained, by purchasers of the chests, when the stock they contain is exhausted.

Letters, containing orders for the chests, which will be sent, carefully packed, to any part of Great-Britain, to be post paid, and a remittance of the value will be expected before the chest is delivered to the carrier.

---

trunk'. According to his brother, John Davy, two arrangements had been prepared for the journey: one measuring 20 in × 7 in × 4 in for 'holiday tests', and the other measuring 12 in × 7.5 in × 6 in and containing necessary instruments (glass tubes, receivers, capsules, blowpipe, a small pneumatic trough, a delicate balance and other necessities).[27] Undoubtedly Faraday, who accompanied Sir Humphry and Lady Davy, would have assisted Davy both with the portable laboratories and in the regular laboratories offered by their hosts in Florence and Rome.

Evidence of everyday applications of the portable laboratory in the second and third decades of the nineteenth century is supplied from the extant catalogues of Frederick Accum and John Newman.[28] Accum himself had arrived in England from Westphalia in 1793 as apprentice to the Brande pharmacy and by 1800 had set up and established his own laboratory and business in Old Compton Street (Soho). From the extensive fifty-four-page

Table 3.2 Range of portable chests from Accum (1817).

CHEMICAL CHESTS

The chemical apparatus and bottles contained in the following Chests are arranged in such a manner, that they might be seen at one view when the chest and drawers are open; they are besides so packed that they can readily be taken out, and when replaced fit in such a way, that the whole, when the chest is locked, may be turned upside down without risk of receiving injury.

Chests, containing a select Collection of Chemical Re-agents or Tests	£3. 3s. to £4. 4s.
Ditto, more compleat; to which are added the Chemical Apparatus usually employed in the processes of analysis by means of Tests	£7. 7s to £8. 18s. 6d.
Pocket Mineralogical Blowpipe Apparatus	£3. to £4.
Mineralogical Travelling Chests	£10. 10s. to £13. 13s.
Accum's Mineralogical Laboratories	£15. 15s to £18. 18s.
The same, more compleat	£26. to £36.
Portable Chemical Laboratories, for carrying on a general Course of Chemical Experiments	£30. to £80.
Chests of Chemical Amusement	£10. 10s. to £18. 18s.
Agricultural Chests	£7. 7s. to £12. 12s.
Cabinets of Chemical Specimens	£21. to £31. 10s.
Chests, containing, in a dry state, the materials for making instantly brisk foaming Soda Water	£2. 2s. to £4. 4s.
Medicine Chests, on an entire new plan, for private families, with a book of directions	£5. 5s. to £21.
Ditto, for the use of the Army or Navy	£20. to £100.

catalogue, with its copious descriptions, it is possible to reveal how chemical practice took place with the aid of the portable laboratory. A summary of items then available is displayed in Table 3.2. In this respect Accum's position, as merchant, analyst, lecturer and popularizer, is none the less important than that of his contemporary Humphry Davy. When Accum departed for his fatherland in 1822, under a cloud of allegations, it seems, others were in the business of supplying the wants of gentlemen-experimenters, including W. H. Potter, who continued Accum's business.[29] Not least in trade prowess at this time was John Newman (then of Lisle Street), who, in 1823, was appointed instrument maker to the Royal Institution.[30] His catalogue is not so extensive as Accum's in chemical matters, although it is possible to detect some development in portable laboratories stemming from researches associated with the Royal Institution.

For W. T. Brande's approach to testing mineral waters, Newman offered a special test chest priced at £7 10s; for Children's translation of Berzelius' study on the blowpipe, there was a special chest costing between £3 10s. and £12 12s.; 'Chemical test chests and apparatus on the plan of Dr Henry' sold for £10; and, most expensive of all, even though thirteen years after Davy's European expedition, there was a chemical cabinet 'according to Sir H. Davy's plan containing a selection of the most useful tests for general experiments', priced between £25 and £35.

3 Are Regular Laboratories Necessary?

Despite a willingness and gentle pressure from the trade to provide for the home experimenter, there was a contrary and deeply ingrained notion that the only rightful place for the conduct of chemistry was within the confines of the regular laboratory. This tradition had to do with the need for providing a suitable room for housing cumbersome laboratory ware, large furnaces and operating by increasingly archaic techniques. Trend-setting examples of what should constitute the regular laboratory had appeared from the time of William Lewis's *Commercium Philosophicum Technicum, or the Philosophical Commerce of the Arts.*[31] Indeed by the time the Aikins published their definitive tomes on laboratory practice, the *Dictionary of Chemistry and Mineralogy*, the broadcast opinion was that 'A convenient well-furnished laboratory is, of course, a principal object with the chemist . . .'.[32] It is not surprising, therefore, to find Faraday later citing, with adulation perhaps, the story of how Alexander Marcet had set about the construction of his laboratory as a matter of utmost priority on arriving at his new Geneva residence.

Faraday himself was so imbued with the importance of the laboratory that he spent no less than fourteen pages of his *Chemical Manipulation* in describing his fortunate circumstance at the Royal Institution to a readership of mechanics for whom he had provided a course on the subject at the London Institution. 'A laboratory', he reminded, 'is a spot where a chemist will pass a great portion of his time.' His wish to go to such lengths may, in part, have been due to his wish to impress upon the minds of aspiring chemists, who just then had acquired a laboratory of their own at the London Mechanics' Institute, that this indeed was the only proper place for chemical practice.[33] In any case, he admits: 'To omit the description of a laboratory in a work devoted to chemical experiment would be erroneous.'[34] Already, in the opening pages, there is projected the now recognizable stereotyped image of the scientist in his proper place of work. Fortunately, something of the tinkering lad who had earned his position at that venerated place through a long 'apprenticeship' under Davy and Brande remained. Elsewhere his lucid descriptions tell of 'ready substitutions', 'contrivances' and 'small, temporary and useful apparatus' which were more in keeping with the practical minds and pockets of his more adventurous readers.

Yet, despite Faraday's reverence for the laboratory, he remained sceptical about those pretentious persons who over-indulged in needless luxuries and, in this respect, he placed a ceiling on the elevation of the laboratory to cathedral status. Consider, for example, the feeling expressed in the following passage:

> Some will not think it [the laboratory] approaches to perfection unless it consists of a large room on the ground floor, well stocked with cupboards, furnaces, and various other etceteras which may be judged to be convenient; and having in connection with a second room, dry comfortable, and fit for the reception of the balance, air-pump and similar apparatus, and a third apartment, which however may be a kitchen, intended to contain movable furnaces, bricks, tiles, sand. . . .[35]

The reference to an extra 'kitchen' with 'movable furnaces' is reminiscent of Berzelius' arrangement in Sweden.[36] However, the mention of 'perfection' and the seekers of 'etceteras' is more likely to have arisen out of social situations created by those closer to home. For example, there were few who would not have admired the fine engraving (drawn by Cornelius Varley) of an ideal research laboratory, complete with several of Knight's furnaces and other 'etceteras', presented in Samuel Parkes's *Chemical Catechism*.[37]

The debate regarding the necessity or non-necessity of a regular laboratory for purposes of study and research had arisen from time to time. Even the Aikins had admitted that Priestley had achieved a good deal with only simple means at his disposal, although this did not detract from their main thrust that regular laboratories were essential places for chemical practice. William Henry, however, in his frequently reissued *Elements of Experimental Chemistry*, had insisted that for study purposes 'A room that is well-lighted, easily ventilated and destitute of any valuable furniture, is all that is absolutely necessary for the purpose . . .'.[38] Accum, too, had no hesitation in pronouncing the traditional laboratory redundant as far as beginners were concerned. Writing under the heading 'Directions for fitting up a laboratory', he explained:

> It was once thought that a regular laboratory, or place built on purpose, and fitted up with forges and brick furnaces, was absolutely necessary for the practice and study of chemistry. This is by no means the case. In proportion as chemical philosophy has been extended, the art of operating has been simplified, new methods of research have been discovered, and new instruments have been added to those which we already possessed. It is no longer necessary to make experiments upon large quantities, for the same properties which characterise minute portions of matter, are also found in a whole mountain of the same substance. Indeed, most experiments *of study* may be more easily performed on a small than on a large scale; and a great deal of expense is saved.[39]

Likewise, J. J. Griffin in Glasgow paraphrased the Henry argument in the early editions of his text *Chemical Recreations*:

> The very prevalent notion that 'a laboratory, fitted up with furnaces, and expensive and complicated instruments, &c &c &c., is absolutely necessary for the performance of chemical experiments,' is exceedingly erroneous. In fact, the truth of the matter is diametrically opposite to this opinion. 'For general and ordinary chemical purposes, (says Dr Henry,) and even for the prosecution of new and important inquiries, very simple means are sufficient: some of the most interesting facts of the science may be exhibited and ascertained, with the aid merely of Florence flasks, of common phials, and of wine glasses.'[40]

The respective target audiences for whom Accum and Griffin had written were very different. By and large, Accum's customers were affluent, while Griffin's readers were mechanics, reluctant to spend and with little time to spare.

The principal technical advance which had thrown into question the design and size of the regular laboratory and simultaneously made possible home and portable laboratories had arisen through the development of small-scale techniques. Telling of this, Accum writes:

> It was by operating upon *grains of matter* that the nature of diamond was established; that no less than four new metals have been detected in the ore of platina; that the composition of the stones which fall from the clouds have been determined; that the metallic basis of the alcalies [sic] have been brought to light; and that the identity of the electrical agency, whether excited by a common machine or by the pile of Volta, has been demonstrated. There is, besides, a degree of neatness gained by operating in the small way in the closet, which is often incompatible with processes conducted on a large scale, amongst the furnaces in the laboratory.[41]

Indeed, experiments performed 'upon large quantities' had naturally developed side by side with the 'regular laboratory', although this whole approach, as Accum related, for 'experiments of study', had fallen into question following the several achievements made by the action of the electric machine, the galvanic battery and the blowpipe on minute portions. From Accum's reference to work on the ore of platinum, it might be deduced that the recent technical achievements of William Hyde Wollaston may have been in mind.

It was certainly the 'neatness' or aesthetic appeal of small-scale techniques which caught Wollaston's attention during a sojourn in Sweden, where he met the Uppsala mineralogist Johann Gottlieb Gahn, a man notably adept in working with small quantities. On returning to England, he developed the art to such a fine degree that Berzelius, who reciprocated courtesies by visiting Wollaston's laboratory at 14 Buckingham Street, Fitzroy Square, reported back to Gahn that the London chemist had reduced the whole of analytical chemistry to a few stoppered bottles from which he extracted reagents in drops.[42] In fact, the whole of Wollaston's laboratory was said to be no more than a wooden tray with handles! Thus, through his sparing approach, Wollaston had paved the way for the removal of chemistry from its expected home, the regular laboratory. A generation or so later, small-scale techniques were recommended by Griffin for the home laboratory, by Ede for his Youth Laboratory and likewise by Faraday under the head of 'minute chemistry'.

4 Ede's Youth Laboratories

The hallmark of what was needed by way of apparatus provision in the 1830s was 'simplicity' and 'cheapness' together with the application of the 'small-scale techniques' just described. This triple requirement was not lost on the Dorking chemist and supplier of toiletries and perfumes to the Royal Household, Robert Best Ede.[43] 'It is a great error', wrote Ede, 'that writers of Elementary Chemistry have accompanied their texts with drawings and apparatus which are, in the first place, very costly and dictating experiments which involve expenses proportionately great. The effect has been to deter thousands from prosecuting science.'[44] With the popular market ripe for development among the attenders at mechanics' institutes and among those in the rising professions, such as medicine and pharmacy, it was Ede who took the initiative to market cheap portable laboratories for distribution on a nationwide scale. In explaining this unprecedented venture, Ede chose to lay stress on Davy's remark, long ago, that chemistry should become an object of study because of its facility with 'simplicity of apparatus' and that its 'implements may be carried in a small trunk'.[45] Hence, throughout 1834 he packaged up 'small trunks' with apparatus and chemicals for accompaniment to the seventh edition of Griffin's *Chemical Recreations*.

These were ready for sale early in 1835, when they retailed for £1. 11s. 6d., or, with stoppered bottles, French polished cabinet, lock and key, for £2. 2s., a little more expensive than stated in Griffin's original announcement. If any anxieties existed in the minds of beginners unsure of chemicals and quantities, then by this time they should have vanished. Even so, there were those who could not reconcile what they had come to expect from public lectures with Ede's cabinets, complaining that they were 'got up on too small a scale'. William Baddeley, a lone voice in the *Mechanics' Magazine*, who had more than once supported the notion of cheap and simple apparatus, countered with the following glowing report:

> I believe that it will be found that the size adopted by Mr. Ede is unquestionably the best size that could be employed. Perhaps neither the size of the apparatus nor the quality of the materials is adapted to the purpose of a public lecturer addressing a large audience, but for private experimental research both are amply

sufficient. . . . The whole is upon a much larger scale than the apparatus which Dr. Wollaston usually employed, and with which he performed many interesting experiments, and made some of the most important discoveries. The fact is, that the chemical properties of a substance characterise equally the smallest portion of that substance, or the greatest mass. That which can be demonstrated of a pound can also be demonstrated of a grain.[46]

There was, of course, much financial benefit to be gained by using small-scale techniques. 'Nay more', adds Baddeley, 'a student in his closet very frequently succeeds in performing an experiment which fails on the lecture-table of the professor; for the accidents which attend the hurry and business of the lecture-room produce unavoidable disappointment.'[47]

Ede had scored a market first, although, according to Baddeley's review, it was not long before his success induced other 'unprincipled persons' to market 'paltry imitations'. Within two years the original plan to provide solely as an accompaniment for *Chemical Recreations* was abandoned when this ran out of print.[48] This caused John Ward, Ede's London supplier, to bring out 'a guide to R. B. Ede's Youth Laboratory' under the title *A Companion: For Footsteps in Experimental Chemistry*. This eightpenny tract, containing seventy-two experiments and sixteen on gases, pre-empted Ede's own publication, *Practical Facts in Chemistry*, which appeared two months later with the author apologetically complaining of 'the obvious haste with which the work has been got up'.[49]

In the second edition of Ede's *Practical Facts of Chemistry* a collection of testimonials, presumably solicited, were prepared for advertisement. Naturally, only the praiseworthy reports appeared in print, although it is worth noting just who was supportive of the plan.[50] Not surprisingly, perhaps, David Boswell Reid in Edinburgh was targeted with an Ede laboratory. He responded: 'I have been for many years in the habit of recommending students of Chemistry to provide Portable Laboratories of their own construction', adding that he thought this selection to be 'very judicious'. Another recipient was Thomas Graham, still at this time Professor of Chemistry at the Andersonian Institution, who said he had 'formed a very high opinion of it' and believed that the student would be 'deeply indebted to Mr. Ede'. Thomas Clark at the Marischal College briefly considered it '*cheap* and very *useful*', while Edward Turner, at University College (London), approved of it 'highly' and wrote that he would have 'great pleasure in showing your Cabinet to my Class, and recommending it to their notice'. The signatories of these testimonials were, of course, those eminent men responsible for organizing the new courses in practical chemistry. In the absence of satisfactory accommodation in their institutions, they might well have seen their personal salvation through Ede's commercial venture.

Finally, the extent of coverage becomes apparent from the newspapers up and down the country who joined in the praise. *The Times* offered laudatory remarks for 'the inventor of this little Workshop' and thought it well-suited

Table 3.3 Full range of Ede's portable laboratories in 1843.

No. 1 Ede's Youth Laboratory
Price 16s
containing more than 40 Chemical Preparations for enabling the Enquiring Youth to perform above 100 Amusing and Interesting Experiments with perfect care and free from danger.

No. 2 Youth Laboratory
Price £1 11 6d
90 Select and Useful Tests, Re-Agents, and Appropriate Apparatus for a course of Entertaining Experiments.

No. 3 Youth Laboratory
Price £2 2 0d
This Laboratory is fitted up with Stoppered Bottles, Improved Spirit Lamp, Lock and Key, and French Polished.

No. 4 Youth Laboratory
Price £3 3 0d
This Laboratory contains additional Apparatus to No. 2 and 3.

No. 5
Price £6 5 0d
130 Select Chemical Preparations and newly-invented apparatus for performing refined Experiments both of demonstration and research in any drawing room, with ease, and success.

No. 6 Chemical Cabinet
Price £8 10 0d
This CABINET is fitted up with extra Apparatus and finished in a superior and more suitable manner, with brass name plate, sunk handles, corners, &c., suitable for travelling.

for 'the aspiring youth, student and amateur'. The *Liverpool Albion* billed it as 'soaring far above any other production of this kind', while the *Liverpool Mail* placed it beyond the boxes hitherto considered as toys and saw that youth may now be led from 'Nature up to Nature's God'. As far as the *Leicester Herald* was concerned, there was simply no place where this 'just celebrity' might not be used. Why not put it 'in the Library of a man of Science, the Studio of the Chemist, or the Boudoir of the Lady amateur' or, indeed, use it 'to render the exhibition of refined experiments even in the Drawing Room'? In East Anglia the *Bury and Suffolk Herald* spoke of 'The march of intellect' and the 'hitherto inaccessible Temple of Science', and even the Church of England *Family Newspaper* elevated it to quintessence by calling it 'the very perfection'![51] Whatever descriptions or metaphors chosen by the reviewers, it is sufficient here to record that Ede's Youth Laboratory, first in combination with Griffin's *Chemical Recreations*, had opened up a vast possibility for study as well as rational recreation.

5 Reid and Griffin: Pioneers of Method and Means

Self-instruction with the aid of portable or other home laboratories, after the style of the Ede–Griffin recommendations, could never be a satisfactory solution for teaching and learning about a practical subject which demanded exemplary teaching. However, few of those who proclaimed the importance of laboratory practice had much idea of how to institute formally run laboratory courses. Nevertheless, in 1829 the Royal College of Surgeons insisted that prospective licentiates should undergo a six-month course in laboratory training. At Edinburgh David Boswell Reid carried out this requirement under Thomas Hope's direction, while London followed suit with Edward Turner offering a three-month course at University College. Three years later, in 1832, the Society of Apothecaries insisted that lecturers should be competent to deal with apparatus and manipulations, and by 1835 they demanded that evidence of such training be given. The pace quickened in 1839, when the medical faculty of London University introduced a new regulation making the laboratory course mandatory. Thomas Graham, by now at University College following Turner's death, charged students £4 extra per term for the necessary practical experience; at King's College J. F. Daniell resisted the regulation, although he eventually capitulated by offering a course in 'chemical handicraft' of the kind to be found in Faraday's *Chemical Manipulation* together with some lectures in metallurgy and domestic economy which, loosely interpreted, were supposed to be befitting of practical chemistry for the all-important attendance certificates, regardless of their actual experience. A further push towards individual practical experience came in 1841, when the Pharmaceutical Society was founded with a teaching laboratory.[52]

It was an important decade for chemistry, as is apparent from the interests of institutions and societies such as the British Association for the Advancement of Science; the Geological Survey and the Mining Record Office, which later became the Museum of Economic Geology (founded 1839); the Royal Agricultural Society (founded 1838); the Chemical Society (founded 1841); the Glasgow Agricultural Society (founded 1843); and the Royal College of Chemistry (founded 1845).[53] From these bodies sprang various moves to promote a scientific education, particularly through the example provided by laboratory chemistry.

Pioneering the new needs became the concerns of two men whose contribution to chemical education has to date received little attention.[54] Both Griffin and Reid were concerned about the need for institutional teaching in well-equipped laboratories and both set about their daunting tasks in the middle 1830s with little else on which to build save their convictions that the job should be done if the masses were to receive any opportunity for individual practical instruction. Griffin, on the encouragement of his friends in the Glasgow Philosophical Society, took it upon himself to design, obtain and supply apparatus on a scale and at prices hitherto unheard of; Reid, on his own initiative, took to running a teaching laboratory of his own design with the freedom to try out his education ideas. Both, in effect, were redirecting their energies after coming unstuck with earlier hopes: Griffin had published some peculiar views on chemical nomenclature which were not particularly welcomed by the hierarchical community of chemists, and Reid had fallen foul of the controlling Edinburgh Municipal authorities with a polemic over the desirability to institute a new chair of Experimental Chemistry at the University.

It is not the purpose here to elaborate on the individual philosophies of these two pioneers, although it is worth observing that their drive to propose solutions to the problem of method and means was directed towards the underpinning goal that chemistry should be taught for intrinsic aims — that is, pertaining to the subject itself — and its educational value — that is, its ability to engage the mind. Both, however, mustered a range of extrinsic aims too, such as those supporting the value of chemistry for its place in domestic economy, luxury and happiness, and theological function. In the present context, however, it was the lack of 'materials of thinking' that occupied their attention as the biggest stumbling block to mass education. Writing of his experiences, Reid explains:

> Considerable difficulty was at first experienced in procuring apparatus sufficiently simple and economical to admit of so large a number of persons as upwards of a hundred young persons operating at the same time; but by taking minute portions of different materials, a mode long familiar to analytical chemists (and particularly well-cultivated by the late Dr. Wollaston) and by using the broad and narrow slips which glaziers separate in cutting window-glass, it was soon found that nothing further was necessary in performing thousands of test and illustrative experiments.[55]

The audience was enormous, with class sizes equivalent to the traditional lecture. Yet he overcame the inherent difficulties of presentation.

The first trials of the 'flat-glass technique' took place over the winter of 1834/5. Two courses operated over twelve meetings: one hundred mechanics in one class and some forty young persons in the other. Something of Reid's approach is apparent from the following extract from a sectional report of the 1835 Dublin meeting of the British Association, at which he induced others to try his method:

> In the mechanics' class, the students were arranged along five boards, each being provided with twenty gas-lamps, one of which was placed alternatively on either side. Each pupil received a blow-lamp, a test-tube, slips of paper on which tests were applied, and also a broad and narrow slip of glass, such as glaziers throw away. The slips were used for the same purposes as the paper, and also for solution, boiling, evaporation, crystallisation, and filtration: the narrow slips, on the other hand, were employed for imitating furnace operations, heat being applied by a common pump or candle, assisted, where this was necessary, by a blow-pipe.[56]

Thus, by the simple expedient of adapting off-cuts of glass, Reid managed to avoid using otherwise costly and cumbersome traditional apparatus.

The general applause for this innovation caused Reid to set aside time from his bread and butter lecturing in order to test further his ideas on the possibilities of practical instruction. Notice of his voluntary exploits in a number of Scottish academies subsequently appeared in a publication of the short-lived Central Society for Education.[57] For those following suit he published an admirable guide, for teachers and pupils alike, in his *Textbook for Students of Chemistry*.[58] The essential accompaniment for this text, perhaps not unexpectedly, was a portable laboratory. Reid explains:

> The small museums of tests and specimens, prepared by Mr. Macfarlane of Edinburgh, are arranged in the manner recommended. They contain about sixty specimens of the important chemical preparations, including a few phials with acids and alkalis, a test tube, slips of flat glass, filtering paper, a test paper. These alone, independent of the larger and more complete portable laboratories and chemical test boxes, cannot fail to be of the highest value to the student of chemistry.[59]

Macfarlane, the Edinburgh druggist, sold Reid's 'small museums' for £2 5s. each and offered a larger portable laboratory for the same purpose at £6 6s.

In parallel with the Edinburgh developments, Griffin in Glasgow set up a commercial venture expressly to supply apparatus for schools. It was, in fact, the first instance in chemistry of the application of the commercial axiom of creating a demand and advance by proffering a cheap supply. As Griffin himself explained:

I have superintended the manufacture, with an endeavour so to combine and organise it, as to reduce the expense of the apparatus to such a degree as to make the introduction of chemical tuition into schools no longer to be dreaded by teachers, as they have hitherto dreaded it, as *a certain course of pecuniary loss.*[60]

Like Reid, Griffin was intimately aware of the expense which even the most willing teachers found off-putting, although he could not tolerate the cheese-paring approach of using glazier's off-cuts, which he derided as 'a last resource for a pinched operator'. Mildly rebuking the Edinburgh doctor, Griffin tells his readers: 'Any kind of glass vessel that has sides as well as bottom is superior to a flat glass plate; and as small tubes can be bought at a penny a piece, a flat glass has hardly even the merit of cheapness to recommend it to adoption.'[61]

As for the principles of the 'chemico-commercial project', Griffin writes:

My study has been, in the first place, to develope [sic] the nature and objects of the more important processes of the science: in the next place, to consider the means and appliances whereby chemists of modern times are accustomed to bring out the result of these processes: and finally, to produce such modifications of those experimental means, as, without abating from their utility, should operate to the reduction of the *three great evils* of *high cost*, *scarcity*, and *difficulty of management.*[62]

Identifying and reducing these 'three great evils' was at the very heart of Griffin's manifesto. Heralding the 'good news' arising from his labours, he continues:

I thought it worthwhile to determine whether these curbing powers prevailed of necessity, or from accident. My investigations have led me to draw the latter conclusion. I find that the evils referred to can be readily overcome, and I shall show . . . that serviceable chemical apparatus can be made as cheap, as plentiful, and as easy to use, as the instruments which have long been employed in illustrating the usual school-taught sciences of geography, geometry, and astronomy.[63]

Thus, by November 1837, the Chemical Museum in Glasgow was able to launch its first catalogue of cheap apparatus sets alongside a full range of Ede's portable laboratories. Among those from whom Griffin had solicited attention was George Birkbeck, who, in a private letter, wrote:

In there providing the chemical student with the means of performing an extensive series of class experiments, you have infused the establishment of the elementary doctrines in the mind, and have thereby provided the best means of extending both the theory and practice of chemistry.[64]

Both Reid and Griffin, the one by example and the other by provision, had urged the route forward towards the concept of a teaching laboratory, not merely for those privileged enough to attend university, but also for the masses of ordinary attenders at mechanics' institutes and schools. In the case of Griffin, no longer did he suggest that an old tea-tray placed before an open window was sufficient for the conduct of chemistry: instead he showed a way forward by describing how a school room might be converted into a laboratory space, with a *pro tem.* iron- or lead-topped worktable (in the place of Dutch glazed tiles, not then available in Scotland) and movable means of ventilation (after the fashion of Berzelius' movable chimney). Nor did he neglect other matters in his embryonic conception of a school laboratory: stone greybeards were considered a reasonable solution to the problem of water supply and waste; shelves and cupboards were envisaged for the storage of bottles, tools and apparatus. Ideally, the worktable was to be divided, with one half fitted with drawers and locked cupboards for containing lucifers and fragile objects best kept away from the prying fingers of youth. Finally, and most importantly, there was to be 'a note book, with pen and ink, for journalising the experiments which are performed'.[65]

Both Griffin and Reid took the 'royal road' south. Reid was the first to arrive in the metropolis, where his principal task was to advise on ventilation for the new House of Commons following the destruction by fire of the old Palace of Westminster. His views on this task, particularly those aspects related to the chemistry of daily life, were considered ideal topics for elementary schools, and, to this end, he lectured before teachers in 1842.[66] However, in 1847 Reid left for Washington to take up a position as sanitary inspector, although not before demonstrating the potential of practical chemistry by teaching to boys at the Borough Road school. No sooner had Reid departed than Griffin arrived to set up a new and grander Chemical Museum in Baker Street, Marylebone.

Throughout the twelve-year development of the Glasgow Chemical Museum, Griffin had not ventured any development of portable laboratories of his own arrangement. In this respect he appears content to have retailed Ede's ranges side by side with his own cheap sets. But with the passing of Ede in 1845 and the inability of Ede's successor to carry on the business successfully, and still in the absence of teaching laboratories, there remained a gap to be filled.[67] It is not surprising, therefore, to find the following announcement in the first *Scientific Circular* to be issued from the new Chemical Museum in May 1849:

In compliance with the request of many of our friends, we are now preparing a Series of **PORTABLE LABORATORIES** of CHEMICAL APPARATUS AND PREPARATIONS adapted for the performance of the Experiments described in *Chemical Recreations.* They will be more or less complete according to the price, but the very smallest will contain materials sufficient for an extensive series of instructive experiments. They will not be toys, containing merely articles adapted for half-a-dozen showy experiments that lead to no results; but they will be so furnished that the student who uses the articles according to the instructions given in the book will acquire considerable experience in chemical researches.[68]

Two months later five laboratories appeared along with the revamped *Chemical Recreations* (ninth edition, 1849), now a text targeted at schoolmasters as well as other practitioners. For them a specially prepared twelve-guinea Schoolmaster's Laboratory 'sufficient for a course of elementary lessons' became available and, on a more lavish scale, a laboratory set at £21 was offered 'for an audience not exceeding one hundred persons' together with sets for individual pupils at £1 or £2.[69]

6 Prospect and Retrospect

Griffin's interest in chemical education around 1850 was matched by his competitors Messrs Horne, Thornthwaite and Wood (successors to Edward Palmer of Newgate Street), and by Knight & Son of Foster Lane. This trade interest had awakened once the Committee of Council on Education (a government body expressly set up in 1839 to provide dispensations for buildings and books) began to allow a two-thirds grant-in-aid towards apparatus provision in elementary schools. This financial help, largely promoted by Rev. Henry Moseley, Britain's first school science inspector, lasted through to the late 1850s, by which time the government backing was given to the 'science schools' of the Department of Science and Art (another bureaucracy set up in the wake of the Great Exhibition to attend to the nation's educational health) in the form of a fifty per cent apparatus grant.

With the motivation of teachers thus controlled by government funding, it might well be expected that the long-standing direct influence of the trade was to wane. Yet the process of encouragements and inducements was slow in action at first, although the movement towards the proper establishment of the school laboratory was inexorable. From the mid-1860s and into the 1870s teachers were encouraged to attend the Department's training courses at South Kensington; from 1872 a further inducement of £1 per student was given to organizing committees who heeded the Department's recommendation to set up a teaching laboratory. At the same time, further monetary reward was offered to teachers who carried out their duties in such laboratories. But, alas, in a system of examinations for which the only regulation regarding practical work was that students ought to be able to describe an experiment, and in a system when 'payment on results' yielded greater financial reward through cramming, there were few who voluntarily took to providing individual laboratory work.

Edward Frankland, who had gained his laboratory experience under Liebig at Giessen in the 1840s, was the author of the 1872 recommendations. Schools were then advised to prepare laboratory space for 20–50 students plus facilities for a preparation room, balance room, library and teacher's laboratory. Later, in 1879, this advice became mandatory, in the following terms:

> For practical work at the Elementary stage it will be sufficient if the laboratory be fitted up for practical work, and be furnished with stout tables, large enough to

accommodate all the students in attendance at the same time, so that each may individually be able to work the experiments set forth in the Syllabus of Inorganic and Organic chemistry; and it must be properly supplied with gas, water, and sinks. *A room which is also used for instruction of the classes of an ordinary elementary day school will not answer the purpose.*[70]

A similar regulation was given for those students taking the Advanced and Honours stages. In their case a 'proper working bench' meant that each student was entitled to bench space measuring 3 ft 6 in by 2 ft 3 in, fitted with drawers, and supplied with a list of apparatus and reagents. Not complying with this regulation would have meant suffering the wrath of the visiting inspector, who might have referred the organizing school committee to that part of the regulation which stipulated: 'no grant will be made unless these requirements at least are absolutely fulfilled'.

Individual practical work in specially designed teaching laboratories became visible from this time.[71] In the interim, from the Griffin–Ede collaboration on portable laboratories, the introduction of Statham's amusement chests, and the Glasgow Chemical Museum's cheap apparatus resulting from Griffin's chemico-commercial project, to the time of Edward Frankland's regulatory demands for those schools and institutes taking the Department's examinations, the trade had indeed directly influenced parents and schoolmasters as well as responding to their several requests. The cases of Wynne and Smithells are examples of the older influence at a time just before they might have benefited from the provision of a school laboratory.

This chapter has covered a period when institutional laboratories were, by and large, non-existent as far as teaching individual practical work was concerned. The trade catered directly for the experimental demands of individuals, whether adult or juvenile, with portable laboratories and amusement chests, even though the long-standing opinion was that chemistry should be practised only within the confines of the regular laboratory. William Henry, Frederick Accum and John J. Griffin resisted this opinion; other trade names, too, such as John Newman, Francis West, Robert B. Ede, William Statham and Edward Palmer, come to the fore as suppliers of home (portable) laboratories. But their makeshift arrangements for autodidactic learning could not, in the end, substitute for the assembly of furnishings and fittings, reagents and chemicals, fume cupboards and service supplies, which commonly constitute that recognized architectural space called the teaching laboratory. Nevertheless, the concept of the portable laboratory as a rational arrangement for amusement or education at home has persisted through to present times with the polystyrene-packaged Open University 'experiment kit'. The era of fine polished deal and mahogany boxes complete with brass locks and keys, however, has long since vanished.

Notes and References

1. Galton (1874).
2. 'William Palmer Wynne (1861–1950)', *Obit. Not. Fell. Roy. Soc.* (1950–51), p. 519.
3. 'Arthur Smithells (1860–1919)', *Obit. Not. Fell. Roy. Soc.* (1939–41), p. 97.
4. William Edward Statham, originally a wholesaler in perfumery, began to trade in 1839 from 23½ High Street, Newington Butts, Southwark, London.
5. Statham (1903).
6. Statham [1862], p. 2.
7. For Accum, see Browne (1925).
8. Accum (1817), p. 46ff.
9. Accum (1819), preface.
10. M. Faraday, *Lecture Notes*, Royal Institution archives, MS F4C, [5]. See also Ironmonger (1958).
11. Newman (1827), p. 16; Cary (1827), p. 16.
12. Anon., *John Bull*, 30 December 1830.
13. The title of this work should not be confused with Griffin (1825).
14. Brougham (1826).
15. Maria Edgeworth's crusade had begun under the influence of her father, Richard L. Edgeworth, a Birmingham 'Lunatic' who, along with Thomas Beddoes, another Lunar Society member, had themselves utilised a child's propensity for play in educating their own children. See, for example, Williams (1968); also Lang (1978).
16. Anon., *Mercantile Journal* (18 February 1831); Edgeworth and Edgeworth (1798).
17. West (1833), p. 7.
18. Joseph, Myers & Company, *Catalogue* in *Classed Catalogue of the Educational Museum*, South Kensington, 1860, and Statham [1862], p. 1, respectively.
19. See, for example, advertisements in texts such as Stephen [1887] and *Illustrated London News*, particularly around Christmastime.
20. Quoted by Gibbs (1951b).
21. Smeaton (1965–6).
22. The translations which diffused this knowledge to Britain were Crönstedt (1770) and Göttling (1791).
23. Weindling (1983).
24. Henry (1806), postscript to preface.
25. Quoted by Farrar *et al.* (1976).
26. For an account of Richard Knight, see Buchanan and Hunt (1984).
27. Davy, J. (1839–40), Vol. 5, 437n.
28. Accum (1817), Newman (1827).
29. Whipple Museum (Cambridge), Trade item B.1.4.
30. Greenaway *et al.* (1971–6), Vol. 5. p. 384.
31. Lewis (1763).
32. Aikin and Aikin (1807), Vol. 2, p. 533.
33. Laboratory illustration in *Mechanics' Magazine* (1826), **8** (26 January 1826), title page.
34. Faraday (1827), p. viii.
35. *Ibid.*, p. 11.
36. For a description of Berzelius' laboratory, see Jorpes (1966), pp. 32ff.
37. Parkes (1822), frontispiece.
38. Henry (1810), p. 35.

39. Accum (1819), p. xxi.
40. Griffin (1825), pp. 73ff.
41. Accum (1819), p. xxi.
42. Although a superb experimentalist, few men were ever invited to Wollaston's laboratory to witness his techniques. His secretive nature is summed up by Walker (1862), p. 49, with the sentence: 'Dr. Wollaston was accustomed to carry out his experiments in the greatest seclusion and with very few instruments.' Brande also displayed some frustration at Wollaston's lack of volunteering information: 'He could never be induced to describe his manipulations in print or to communicate to the world his happy contrivances', although, he indicated, 'we owe to his numerous abbreviations of the tedious processes a variety of improvements . . . which have gradually become public property'. Brande (1848), p. cii.
43. For Robert Best Ede (c.1799–1845), see John Atlee, *Reminiscences of Old Dorking*, Typescript, Education Department, County Library, Guildford, 1912.
44. Ede (1837), p. viii.
45. Davy, quoted by Ede, *ibid*.
46. Baddeley (1835), p. 163. (N.B.: Baddeley's own campaign for cheap apparatus commenced following Brougham (1825).)
47. Baddeley (1835), p. 164.
48. Ede's reference to 'the obvious haste' and 'the unavoidable circumstance' would seem to allude to the fact that the 1834 edition of Griffin (1825) was out of print. See prefatorial remarks in Griffin (1858).
49. Ede (1837), prefatorial remarks.
50. Ede (1843), endpapers.
51. Quoted in *ibid*.
52. The Pharmaceutical Society laboratory catered for eighteen students in 1845. For details of this laboratory see *Pharm. J.* (1846), **5**, 314–24.
53. For an excellent account of these developments, see Roberts (1976); and for the development of the discipline of chemistry, Bud and Roberts (1985).
54. I am currently investigating J. J. Griffin's role in both business and educational contexts; for an outline of D. B. Reid, see Layton (1975).
55. Central Society for Education, *First Publication*, 1837, p. 68.
56. Reid (1835b).
57. 'Notice of a system proposed for introducing chemistry as a branch of elementary education', in Central Society, *op. cit.* [note 52].
58. Reid (1835a).
59. Chambers' *Edinburgh Journal* (1836), **5**, 158. *Note*: J. F. Macfarlane, of 17 North Road, Edinburgh, held an appointment as chemist and druggist to Her Majesty: Edinburgh *Directories*, 1836–42.
60. Griffin (1838), p. ix.
61. *Ibid.*, endpapers.
62. *Ibid*.
63. *Ibid*.
64. Birkbeck to Griffin (copy in Birkbeck's hand): Letter 65133. Archives of the Wellcome Institute for the History of Medicine (London).
65. *Ibid*.
66. Reid (1844).
67. Atlee, *op. cit.* [note 43], p. 17. There are, however, signs in the preface to Griffin (1849), dated 15 December 1846, that he was giving the matter consideration prior to leaving Glasgow.
68. Griffin's *Scientific Circular*, No. 1 (May 1849), p. 4.
69. Griffin's *Scientific Circular*, No. 3 (July 1849), pp. 39, 42.
70. Department of Science and Art, *Directory* (1879), p. 44.

71. The regulations for physics laboratories, not explored here, were less severe and only became a strong recommendation in 1899, when a room, free from vibrations, with blinds to exclude light, rails from which to hang objects, and plenty of light, plus a guide on furnishings, became the basis for architectural design: Department of Science and Art, *Directory* (1899), pp. 121ff.

Section 2

THE EXTENSION OF

LABORATORIES TO PHYSICS

4

History in the Laboratory: Can We Tell What Really Went on?

David Gooding

The idea of places set apart for experiment was well established by the early nineteenth century. Then, as now, some types of experiment could be made only at the site of special equipment, while others could be made only in unusually large spaces.[1] Still others, such as the mapping of geomagnetic phenomena or testing the effectiveness of a vaccine, could not be made within the confines of a building at all.[2] The legitimacy of reducing the scale of natural phenomena to get them inside the laboratory and of reproducing them by artificial means was still contested during the last century.[3] None the less, natural science — already associated with 'the laboratory' — became increasingly identified with it during the nineteenth century. The importance of laboratory work to the development of science, to deciding technical issues in legal proceedings, to industry and to national security was widely accepted. The nation needed science and scientists needed places to conduct experiments.[4] A book about such places would not be complete without an account of experimentation in them. This chapter addresses the recovery of experimental *activity* rather than the social, epistemological, architectural or other characteristics of the space in which it took place. My examples are a public trial by John Herschel in the lecture theatre of the London Institution in 1823 and some private, exploratory work done by Faraday in the basement laboratory in which he assisted Humphry Davy and Thomas Brande at the Royal Institution in 1821. The first example illustrates the variable identity (or plasticity) of experiments and shows that experiments may be made inside laboratories but their historical identities are made elsewhere. The second example shows the importance of understanding the procedural and manipulative aspect of observation and experiment. Skills and techniques are essential to science and technology. If these could be recovered from written sources or inferred from apparatus, then the usual historical methods would suffice. I argue that since skills cannot be recovered from the familiar literary and material forms of evidence — manuscripts, publications and instruments

— historians of science should, if possible, venture beyond these, to study the activity that produced them.

For experimenters, the problem with experiments is that they rarely work according to plan, if they work at all. For historians, the problem with experiments is that scientists' accounts of them naturally reflect the plan or the finished product, rather than actual practice. The discrepancies between practice and accounts of it are not insignificant. The procedural and technical side of experiment is essential to the development of new science.[5] Experimental practice is also important to the presentation and defence of results. When responding to criticisms of their experimental methods, of inferences and interpretations, and even of arguments which spell out the theoretical significance of a result, scientists often rely as much on what they have learned through experimenting as on what they have gained from an intellectual understanding of the theoretical context of their work.[6] Recent studies of controversies involving experiment show that almost any aspect of an experiment can be scrutinized and challenged — including 'know-how'.[7] Empirical results are never completely independent of the practices that produce them. Facts are practice-laden as well as theory-laden.[8] Of course, we can understand much of the meaning of research reports and justificatory arguments and we can treat laboratory notebooks and published treatises as a purely literary genre. However, I doubt that we can explain what their persuasiveness has to do with *experiment* if its preparatory, enabling role is ignored.

The grounds for belief in experimental claims are usually laid in the private context of a laboratory.[9] This means that we need to consider differences between the account of experiment and the activity it records. These are greatest in published reports, whose purpose is to persuade others of the existence, relevance or importance of a result, not to record what went on in the laboratory.[10] Work in present-day laboratories is accessible — anthropologists of the laboratory observe it as it goes on. Historians must reconstruct experiments from private records such as laboratory notebooks and correspondence. Private records are less rhetorical in character than published accounts, yet they already contain an element of reconstruction and they are incomplete. There are two sorts of incompleteness in the recording of experiments. The first, and most obvious, is that there are gaps: nothing was written down for the afternoon of (say) 2 September 1821, or for the whole of that month or year. In the second section I look at John Herschel's reconstructions of an important experiment, of which he himself had kept no records. The remainder of this chapter deals with a second sort of incompleteness, the omission of what Polanyi called *tacit knowledge*: the skills, techniques, assumptions of which practitioners were either unaware, or which, by their very nature, could not be recorded in writing or drawings.[11] If much of what experimentalists do cannot be recorded, it cannot be recovered by reading texts or even by studying apparatus. Well-documented examples of experimental work — for example, in manuscript notebooks of Newton,

Lavoisier or Faraday — suffer from incompleteness of both kinds, and the majority of scientists left no private records of their experiments at all.[12] It would appear that the pre-articulate and procedural aspect of experimentation means that its history cannot be written with the same confidence as, say, the history of scientific ideas and scientific institutions or the accounts of contemporary science given by contemporary observers of what really goes on in laboratories.[13]

I have just indicated why the history of the published experiment is not the only history that matters. The history of experimentation that we have is largely based on work reported in a public context of argumentation rather than the context of exploration and invention. This is mainly due to the belief that experiment has always been primarily a form of empirical check on theories. Thus, experiment usually appears in its finished and perfected form as demonstrating a phenomenon or as enabling scientists to choose rationally between rival theories. Until recently science's history has been primarily about ideas, individuals and institutions, and it has been quarried from written sources. However, the less literary view of science now emerging does not portray experiment as the handmaiden of theory, to be approached through the history of theories. Experimental grappling with the world involves skills as well as plans, instruments and concepts. The one-dimensional view of experiment as empirical check now appears as an artefact, largely of mid-nineteenth-century theories of scientific method (such as William Whewell's).[14] This raises a question for historians of science. Can the history of *experimental* science be written if we never go behind the reconstructions of public (or even private) accounts, to find out what went on in laboratories and workshops? If we do not, are we not confined to writing the history of the *literature* of experiment? In that case, it could be argued, 'the experimental event' reduces to the dissemination of a text and the history of laboratories becomes the history of places set apart for the production of special kinds of discourse.

1 Recovering Experiment

The history of experimental science need not be based solely on written sources. Other methods can take us through the gaps in these sources, beyond texts and into the procedures they imperfectly record. These methods are experimental, but not in the way that natural science experiments are. A changed image of experiment reflects a different view of history of science in relation to its subject: the historian repeating experiments is trying to recover the 'how' of past science, not to replicate its results. This is an important change, because the idea of experiment as a deliberate, logical process that gives decisive answers to questions about nature is difficult to doubt as long as it is taken for granted that more recent science comes closer to the Truth about Nature. On this assumption, scientists' accounts — and interpretations by historians — can be supplemented and corrected in the light of scientific progress; gaps in the documentary evidence can be filled by more recent and more precise (or correct) experimental outcomes, or more recent interpretations of their significance. Assuming that failed experiments were due to shortcomings in the apparatus, historians who did repeat experiments and reproduce instruments tended to focus on instrumentation as a technological constraint on theorizing. Their main interest in repeating experiments was to assess their sensitivity or precision and to evaluate contemporary claims about these.[15]

The history of science is littered with experiments which failed to detect effects later discovered with more accurate, more powerful or more sensitive instruments. The new view of experiment highlights the processes, so that anticipation and precision are less important than understanding how scientists learn from experiments. I do not mean to suggest that what they learn are bold, self-evident facts — rather, they learn how to conduct and improve the experiments as means of interacting with nature so as to elicit and disclose natural phenomena. Theories show what experiments need to be made, but they rarely show how to do them. This must be learned as experiments proceed.[16] Experimentation has a constructive, inventive aspect which, as I pointed out earlier, plays an important enabling role in the presentation, interpretation and subsequent critical scrutiny of experiments.

The idea that historians can repeat experiments to recover more information about them is not new. Belloni repeated anatomical observations in order to interpret the otherwise unintelligible observational reports of seventeenth- and eighteenth-century anatomists.[17] Hackmann has examined eighteenth-century electrical experiments through detailed studies of contemporary apparatus and instruments.[18] Settle repeated some experiments of Galileo's on falling bodies and inclined planes. He evaluated the technical feasibility of the latter but was able to identify experimenter-effects in the falling bodies experiments. These effects may explain why Galileo believed that he saw a lighter body fall more rapidly (at first) than a heavy one. Settle showed that it is difficult to ensure that an experimenter drops two balls simultaneously, because the difference in their weight affects his ability to judge the simultaneity of action of left and right hands letting go.[19] Modern, more precise methods of observing the experimenter brought this experimenter-effect to light. What is important here is not just their modernity or precision, but the attention to the manipulation of objects and processes. The second of my examples (below) shows how recovering something of the process of an experiment by repeating it can improve upon the very limited understanding of procedures gleaned from textual sources alone. Earlier I mentioned that the tacit nature of skills makes them invisible. I turn now to another obstacle to recovering experiments.

The Plasticity of Experiment

In order to recover the process of an experiment, we need to make the enabling role of skills visible. To do this we must be able to recover something of the changing understanding of the nature and significance of the experiment.[20] One feature of experiments closely connected to learning is their *variability*: scientists perform them over and over again with modifications which may be systematic and intentional or intuitive and exploratory. Variation is one of the key factors in the success of experiment. Frequency of variation in Faraday's experiments is apparent from his laboratory notebooks: these records suggest a lot of unrecorded and unpremeditated variation. This introduces uncertainty into our interpretation of what 'really' went on. It is helpful to distinguish between deliberate variation and the variability induced by a changing understanding of an experiment. I call the latter the *plasticity* of experiment. Plasticity refers to the variable identity of an experiment and has two sources: one is the process of developing a working understanding of experiment; the other is that subsequent changes in understanding the meaning or significance of an apparently finished experiment are brought about by later events, such as criticism of its performance or controversy about its significance, the outcomes of later experiments and new theoretical developments.[21] The second of my examples shows the changing identity of an experiment in process, while the first example illustrates retrospective plasticity — the reconstruction of a 'finished' experiment.

2 The Changing Identity of an Experiment

On 8 November 1845 *The Athenaeum* began its 'Weekly Gossip' column with the report that Mr Faraday had announced 'a very remarkable discovery . . . that a beam of polarized light is deflected by the electric current, so that it may be made to rotate between the poles of a magnet . . .'.[22] Faraday had in fact rotated the plane of polarization of a beam of light by use of a magnetic field. This is now known as the magneto-optic or Faraday effect. Among the letters of congratulation Faraday received was one from John Herschel dated 9 November. Herschel reminded Faraday that 'a great many years ago . . . I tried to bring this [same connection between electricity, magnetism and light] to the test of experiment . . . (I think it was between 1822 and 1825)'. The fact that Herschel could not even recall the year is significant, in view of his later recollections of this experiment. Herschel tried to reconstruct the event with the help of friends. To find out who had been present, he wrote to William Hasledine Pepys.[23] Part of this letter is shown in Figure 4.1. Pepys had been secretary of the Royal Institution and had in 1823 invented a powerful 'quantity' (or high-current) battery which Herschel had been keen to use for his experiment. It is interesting to note that, although Herschel remembered very little about the experiment, he could recall the unusual structure of the

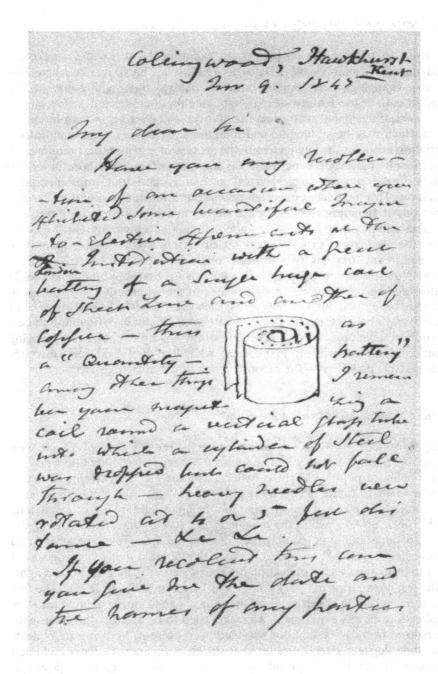

Figure 4.1 The first page of John Herschel's letter of 9 November 1845 to William Hasledine Pepys. From Royal Institution Archives, Pepys papers, courtesy of Dr John Squire.

element of Pepys's battery, and included a sketch of this in his letter, as a mnemonic. Pepys replied with a list of those he thought had been present.[24] He seemed to be more concerned with who had witnessed the event than with how the experiment had been done. Pepys listed many of the most active chemists and electrical philosophers of the day, including Humphry Davy, W. H. Wollaston, Charles Babbage, J. F. Daniell and Faraday. What drew them to the London Institution on 14 March 1823 was not, apparently, Herschel's theoretically important experiment, but a desire to see a display of electromagnetic effects produced with Pepys's new 'magnetic coil' or quantity battery. Pepys recalled in his reply to Herschel that although 'you were desirous of trying some experiments on light . . . being so taken up at that time prevented me paying the attention I could have wished to you'. It appears that, despite its importance to the shape of physical theory, Herschel's experiment had had to wait until the very end of the display, when the battery was all but exhausted. So a new, powerful experimental battery was not used to best effect, because people were actually more interested in something else.

It would be wrong to suppose that Herschel's attempt to discover a new connection between physical forces should have taken precedence over a display of the power of the battery. The purpose of the occasion was to exhibit it as a new demonstration device. Demonstration and display have always been important to scientists. Electrical and magnetic displays were a livelihood for many members of the London electromagnetic network; for others they were a means of winning public and government interest in, and support for, the new field of electromagnetism.[25] This does not explain why Herschel failed to repeat the experiment on another occasion, except in so far as it suggests that testing theoretical views was just one of the many things that occupied his attention, until he read the report of Faraday's success in November 1845. It is interesting to see how Herschel's view of his 1823 experiment now changed.

At first he assumed that Faraday's success vindicated his own earlier views. This would mean that he had realized in 1823 all but one of the conditions necessary for Faraday's recent success. Believing this, he seized upon Pepys's confirmation of his recollection that his experiment had come at the end of the session: surely it would have succeeded, if the experiment had been performed with the battery in its fully charged state. On the other hand, Herschel also (implicitly) acknowledged his failure to observe the methodological requirement that experiments be varied. Despite his continuing interest in the interaction of light and matter. Herschel had not repeated his 1823 test, let alone made the series of 'cross-examining' experiments required by the methodology propounded in his *Preliminary Discourse* of 1830.[26] There he insisted on the need for experiments designed to check or cross-examine the results of other experiments. This ruled out any suggestion that he had anticipated Faraday. In a postscript to his letter to Pepys, Herschel insisted that 'he who proves [experimentally] is the discoverer'. This places experimental success above unconfirmed theoretical anticipation, acknowledging that even though experimental results may vindicate theoretical predictions, such theories often fail to anticipate the conditions necessary for experimental success. Whenever this gap between theory and experiment is overlooked in the wake of a success, the result seems to have leapt directly from a necessary convergence of prediction and practice. In fact, this convergence requires a systematic and lengthy process of varying the experiment, during which the conditions necessary for the outcome have been learned.[27]

By January of 1846, Herschel's interest in the forgotten 1823 experiment had developed further. He viewed the failed experiment which he had hardly been able to recall two months earlier as providing positive evidence for an alternative to Faraday's provisional interpretation of the magneto-optic effect. This came about when, on learning from Faraday of the real nature of the discovery, he realized that he had not anticipated him at all. Faraday had begun by looking for an optical effect of an *electrostatic* field (not a magnetic one) and had then varied this experiment in two crucially important ways. He substituted a magnetic field for the electric one and he introduced an optically dense medium into the field (borate of lead glass). He observed the effect he sought only when this was present.[28] Faraday had tried the glass as part of a long series of variations, and not — as far as I can tell — because he had theoretical grounds for supposing that a dense medium is necessary. In fact, he tried to get a result in air (thus 'repeating' Herschel's experiment) and believed that the effect should occur even in a vacuum.

Herschel had theoretical grounds for disagreeing. Thinking that the effect could occur only if matter is present, he soon presented his result of 1823 quite differently: not as a failure to anticipate Faraday's 1845 experiment but as a piece of evidence in its own right. In this guise it would suggest that the effect cannot take place unless the field is occupied by matter. Of course, this meant reinterpreting the fact that had at first been so important to Herschel, that the null result had been due to the tired state of Pepys's battery. As he explained to Robert Hunt: 'Had the Magnetic Helix acted *directly* on the polarized ray, my experiment prepared in 1823 . . . would have shewn it.'[29] In the course of only two months, his forgotten experiment had acquired a new life, without so much as a single attempt at repetition by Herschel himself. Believing his and Faraday's results showed that magnetism does not act on light directly but through its effect on the medium through which it is passing, he persuaded Hunt to undertake further experiments.

Variation and the Plasticity of Experiment

Herschel's practice did not live up to his methodology. To suppose that this is merely an example of bad practice is to miss the real significance of this episode, which shows that the identity of an experiment — its importance and significance — is not fixed: it is plastic. Herschel's reconstruction of his own experiment is not untypical, nor is it surprising. Experiments are highly contextualized, rather like premises in an ongoing argument. During an experimental controversy the relative importance of — and even the meaning of — individual statements (or pieces of evidence) may alter in response to points singled out for criticism by the other side. Here the weakened state of the battery (previously important) became unimportant because of the theoretical significance of the lack of an effect in air. Other examples of experiments whose identities changed over an extended period of time include those made by Robert Boyle with the air-pump, with which he established experimental practices for the new mechanical philosophy. Shapin and Schaffer have shown that there was never a definitive version of this instrument.[30] The pump was continually revised during the 1660s in response to experimental failures and the criticisms of opponents of the new experimental philosophy. Swenson's study of the ether drift experiments between 1880 and 1930 suggests that the same was true of Michelson and Morley's experimental programme. The interferometer was repeatedly revised and improved; even so, the experiment was never performed to meet Michelson's own specifications and expectations.[31]

3 Faraday's Investigation of Electromagnetism

The previous section described Herschel creating a history for an experiment, first when he learned what its outcome should have been and then in response to his and Faraday's ideas about its physical significance. Historians of science are always in this position. Yet it is possible to recover something of the situation of a scientist recording what he was doing, *before* he knew what the outcome should have been or what it might mean, by supplementing written sources with practical knowledge recovered through the repetition of experiments.[32] Strictly speaking, the distillation of an experiment to a quintessential or 'actual' discovery sequence, familiar from textbook accounts, is an ephemera. Since the route is clear only after the terrain has been explored and mapped, it cannot have been *the* route, foreseen from the beginning. Plausible discovery sequences are not inaccessible, however, as I hope the next example shows.

The notes recording one of Faraday's studies of interactions between magnetized needles and current-carrying wires are shown in Figures 4.2 and 4.3.[33] These notes were written during one of Faraday's earliest investigations of electromagnetism. I shall interpret them with the aid of recent attempts to reproduce some of his observations. The purpose was to improve the interpretative use of literary evidence about material practices. Specifically, we hoped to refine the interpretation of these manuscripts, using practical knowledge recovered (or made explicit) by the repetition.[34] A further stage of the project (to be discussed elsewhere) is to compare the published account of these experiments with the record shown here.[35]

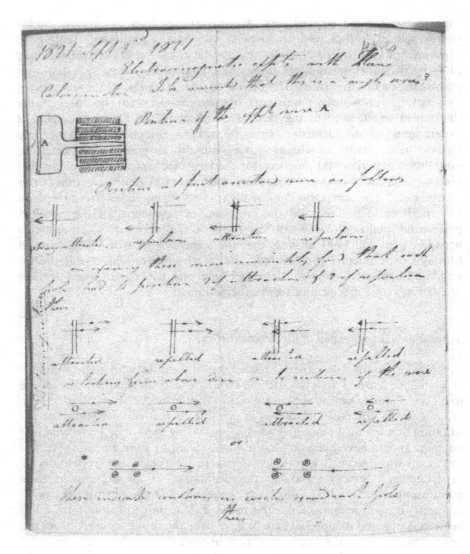

Figure 4.2 (above and opposite) The first two pages of Faraday's notebook record of experiments conducted on 3 September 1821. From Royal Institution Archives, Faraday papers.

3 September 1821

The experiments examined here were part of a study of the interactions of magnets (or magnetized needles) and wires carrying electric currents. We need not assume that Faraday was testing well-articulated expectations in the way that standard discovery accounts lead us to expect: pushing our expecta-

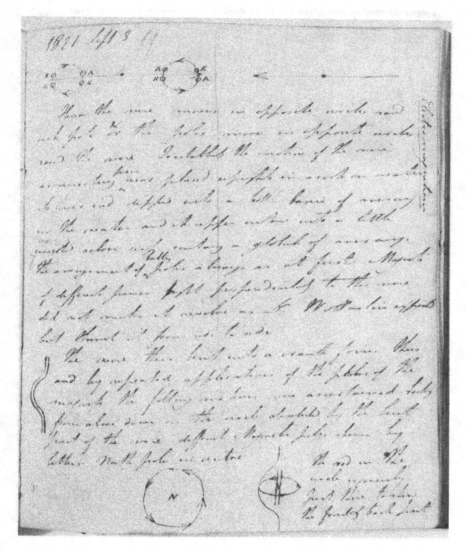

tions aside, we can see that Faraday repeatedly modified his expectations as he experimented. He had several possible interpretations of these and was interested in resolving a tension between two observed properties of the interactions: one concerned attraction and repulsion — namely, the fact that whether a needle is attracted or repelled seems to depend simply on position. Part of his problem was to interpret this fact in a way that made it compatible with the other main anomaly: no known force exhibited structure in the way electromagnetism did, as circles around things rather than attractions or repulsions along straight lines drawn between things. He was interested in both attractions and repulsions *and* in structuring effects, or alignments.

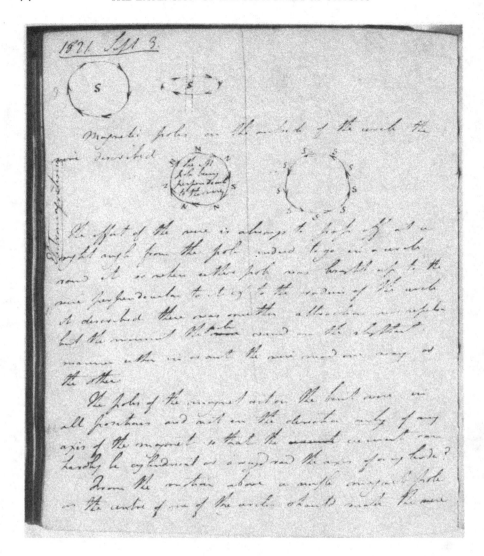

Figure 4.3 (above and opposite) The third and fourth pages of Faraday's notebook entry for 3 September 1821. From Royal Institution Archives, Faraday papers.

According to his record, Faraday first positioned the experimental wire vertically rather than horizontally, so that it could be moved more freely round a vertical wire. He then recorded the tendencies of the needle to approach or recede from the wire, representing these visually, as properties of position. He then made a 'more minute' observation of the behaviour of the needle from above and from the side. This suggested that each pole of the needle had not one, but two positions of attraction and repulsion. Finally, he

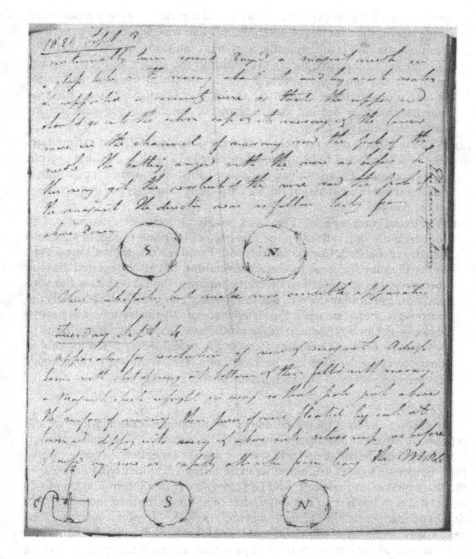

gave a pictorial summary which compressed all of the preceding results into a single pair of diagrams (shown in Figure 4.2 at the top of the right-hand page).

When we tried to repeat these experiments, we found out several things we could not have learned from the manuscript or published accounts:

(1) It is extremely difficult to get a needle to remain still near a current. The needle twists, owing to tension in the suspension; it is pushed about by air currents; it interacts with the earth's field; and so on. The biggest problem is that it tends to touch the wire (this immediately alters its magnetic

properties). In these experiments we found it easier to allow the needle to settle down, then move the wire around. The wire was placed in various positions and the resulting attractions and repulsions were recorded. We did not get alignments, nicely ordered into the circles he drew, leaping ready-made from the laboratory bench. Although Faraday had mentioned some of these difficulties in his 'Historical Sketch of Electromagnetism', they are not recorded in the notebooks subsequently incorporated into the *Diary*.

(2) We found that it does not behave as if the poles are at its two extremities. The needle appears to be attracted or repelled at points a little way from its extremities. Faraday does not record this, although he could easily have verified this behaviour (as we did), using iron filings to map the field of the needle.

(3) We found that one of Faraday's diagrams must be incorrect. The sequence of observations on the first page of his notes in Figure 4.2 could not be reproduced. Comparing this with Faraday's diagrams allowed us to verify Faraday's representational conventions, but also suggested that the top arrow in the *third* diagram of the first row should be below the wire and not, as Faraday drew it, above the wire. Here (as with Herschel) the purpose is not to find fault: the significance of this slip is that it was easy to make — this brings home to us the problems raised by the unfamiliarity of what any observer (including Faraday) is experiencing when looking at new phenomena. One of the most important is how to make one's experience visible — and memorable — through sketches.

To sum up our findings on Faraday's 'more minute' examination, we did reproduce observations resembling those he recorded, but only after (i) learning to manipulate the wire in ways he does not mention, (ii) determining that the needle's poles are not at its extremities, (iii) identifying an error in the diagrams on his first page, and (iv) revising our first interpretation of this error: if it is in the *second* drawing rather than the third, then the two sets of drawings are consistent. Finally, having reached this point, we could be reasonably confident of the change of frame of reference needed to interpret the diagrams at the bottom of the first page and the top of the second. (Rotate the whole through 90° about an axis perpendicular to the wire and parallel to the needles.) Most important, however, is the difficulty of *seeing* what Faraday records that he saw, even though we knew what we were supposed to observe. This shows that expectations do not produce experiences: Faraday's observations must have required a lot of practice. In science, as in the visual arts, a great deal of skill underlies the experiences that observers have.[36]

The next part of Faraday's account shows that he began to refine the experiment. How did he reach the prediction that 'The [effects] indicate motions in circles round each pole ...'? Here his verbal description is accompanied by a diagram of the apparent motions. This interpretation could be expressed pictorially and verbally: 'Hence the wire moves in opposite circles around each pole and/or the poles move in opposite circles round the wire.' Faraday could now imagine these motions, but had not yet tried to produce them. His verbal description of the needle's behaviour is false. He could not 'indicate motions in circles round each pole' with the arrangement drawn here, because a needle does not approach or recede from a wire if, as shown in the diagram, the two are in the same plane.

Here it is tempting to interpret the diagram in terms of what we now know (thanks to what Faraday found out), underwriting the plausibility of our reading by the assumption that Faraday wanted to get on with testing predictions. Repeating these experiments helps set this hindsight aside. Then it is just as plausible to read the diagram as a 'summary' of the preceding operations and their outcomes. Read this way, it makes an inference from the temporal sequence of operations to an ordering or organization of their outcomes, tacitly transforming the temporal sequence into a spatial arrangement. Since the needle and wire must be in the same vertical plane, Faraday imagines that the *wire* describes circles instead of the needle. Here our experience with the difficulty of rendering observations visually suggested that this change may have been stimulated by a practical problem of representation. Faraday had to record and describe in two dimensions an interpretation of something that he imagined as happening in three dimensions.

Faraday now had the idea that the wire can be made to exhibit some sort of circular motion. This suggested a new configuration of wire and magnet. He placed a piece of wire so that 'its lower end dipped into a little basin of mercury in the water and its upper entered into a little inverted cup containing a globule of mercury' and brought a magnet 'perpendicularly to this wire'. At this stage Faraday has the notion that the wire should move, and records that magnets 'did not make it revolve as Dr. Wollaston expected, but thrust it from side to side' (see Figure 4.2). This suggested that the wire would move bodily, not on its axis, so a further modification of the set-up was called for. By further exploration he resolved this into another geometrical relationship — the wire tends to move in either direction along a line perpendicular to the axis of an approaching magnet. This enabled an important variation: he bent the wire into a crank form 'and by repeated applications of the poles of the magnets, the following motions were ascertained, looking from above down on the circle described by the bent part of the wire . . .'. A subsequent repetition of this enabled us to work out what these 'repeated applications' of the magnet would have been.[37] He brought the magnet towards the 'centre of motion' of the crank — that is, its axis. We surmise that the crank swung around so as to strike the magnet and that, by moving the magnet quickly out of the way and reinserting it behind the wire, he got the crank to keep moving. These operations do produce an irregular motion which, with practice, resembles what he recorded in the *Diary* as a ring, where the ring represents the circle described by a point on the rotating crank and the 'repeated applications of the poles' are indicated by the letters 'N' and 'S' around the rings in Figure 4.3.

Faraday now knew that a wire crank would try to rotate, as it were, around the *pole* of the magnet, but he had not yet realized the possibility implied by his circles — that is, *continuous* motion of a wire about a fixed magnet. The motion could not be continuous, because the pole had to be close to the *axis* of the crank (not outside the circles in Figure 4.3). To obtain a continuous motion, he had to move the magnet out of the way and reintroduce it as the crank went by. He established that the action occurs if the magnet is outside the plane of the circle shown in the figure. (This result would later be important to a physical interpretation in his controversy with Ampère, because it showed that the magnet was not necessarily interacting with the current construed as an entity confined to the axis of the wire.[38]) This enabled him to infer 'from the motion [of the crank] above' that 'a single magnet pole in the centre of one of the circles should make the wire continually turn around'. Faraday now saw how to construct an apparatus that could realize continuous circular motions of the wire. This was the first time that anyone had produced continuous motion of matter by converting the chemical energy in the battery into electrical and then magnetic forces.

Here, again, the brevity of the laboratory record is misleading. Our repetition of this part of the experiment shows that there was still a certain amount of trial-and-error adjustment to be made to this prototype before a continuous electromagnetic rotation could be seen. At this stage the effect was still personal to Faraday. No one else had seen it; nor could they see it without instruction from him. This is why his concluding entry for the day in Figure 4.3 read: 'Very satisfactory, but make a more sensible apparatus.' This shows how quickly he turned to the problem of demonstrating the circular motion. To make the phenomenon real, he had to make it public — that is, accessible to others. To Faraday this was a problem which would be solved more effectively by practical means than by verbal argument alone. He made a small, pocket-sized apparatus to send to scientific colleagues, thus making it unnecessary for them to acquire the tacit skills that he had built up.[39] Because the rotation apparatus embodied these skills, it bypassed the problems and pitfalls he had encountered which would otherwise delay the replication of his result by others.

Learning to do these experiments involved many material and conceptual manipulations of objects. Successive interpretations underwent considerable clarification in experiment in order that the material conditions necessary to realize the effect could be learned. This illustrates the non-retrospective sort of plasticity that I introduced earlier, which is characteristic of the microstructure of learning in close interaction with nature. The identity of this experiment continued to change even after its conclusion and publication, because of factors extrinsic to the experiment itself. It was first construed as showing motions no different from some that Wollaston and Davy had recently failed to obtain (making it appear that Faraday — still a very junior figure in the world of metropolitan science — had either plagiarized Wollaston's experiments or trespassed upon the work of his seniors). However, Wollaston's somewhat ambiguous ideas were not clarified in his experiments in the same way that we have seen Faraday refine and clarify his. The distinction between the motions Faraday produced and those Wollaston may have intended became clear only when Ampère claimed to have made a current-carrying wire rotate on its own axis under magnetic influence.[40] Far from confirming a prediction of Wollaston's, Ampère's result made it possible to discriminate between his failed experiment and Faraday's successful one by attributing to the former the natural phenomenon that it could have produced — just as Faraday's experiments of 1845, showing magneto-optic rotation with glass and the lack of it in air, led to retrospective changes in the identity of Herschel's 1823 experiment. The identity of these experiments was not fixed once and for all by their public outcomes. This shows why we need to venture inside the laboratory.

4 Conclusion

For historians the recovery of experiments must often end with the study of surviving manuscript accounts. I hope I have shown that reproducing experiments can tell us quite a bit more than written records alone can. I noted earlier that the difficulty of seeing what Faraday 'says' that he saw indicates that his observations depended upon a lot of practice. Another important example is that making induction rings to repeat Faraday's 1831 experiments on electromagnetic induction established that it takes several days and several pairs of hands to wind an induction ring as Faraday did.[41] This finding shows how misleading the laboratory notes can be. The famous entry for 29 August 1831 begins with trials on coils already prepared. Although we know that time and effort had gone into these preparations, the record gives no indication of just how much was needed to begin these experiments.[42] Showing how much time must have been invested prior to 29 August casts light on a different aspect of experiment. It suggests far more prior commitment to the 1831 experiment — indicating that Faraday had stronger expectations about its outcome — than we can surmise for the 1821 rotation experiments. The latter required far less preparation but rather more manipulative skill. Different aspects of experiment such as these can be recovered by repeating them. As clues to the material and practical context of the experiments, they are valuable aids to the interpretation of scientists' written accounts about what they do in the laboratories.

Acknowledgements

I am grateful to Jane Leigh and Susan Perrett (then of Bristol University) for exploring Faraday's electromagnetic experiments with such ingenuity and enthusiasm, and to Bill Coates, Frank James and Bryson Gore for help with the 1821 rotation experiments. I thank the librarians and staff of the Royal Society and the Royal Institution for their assistance, the Royal Institution for permission to use material from the Faraday Collection and for assistance with photographs, and Dr John Squire and the Royal Institution for permission to use material from the Pepys Collection. Earlier versions of this chapter were read at seminars in the History Department, Simon Fraser University, and to a symposium on *Laboratories* at the Royal Institution in 1986, and to the Humanities Department at Imperial College, London, in 1987. I should like to thank several participants for their comments, especially Hannah Gay, Beryl Hartley, Frank James and Jim Secord. I have also benefited from discussions with Trevor Pinch and Simon Schaffer.

Notes and References

1. For example, Faraday's small-scale experiments on electrostatic induction led him to examine the electrostatic properties of a large space enclosed within a much larger space — thus, the first Faraday cage was built in the large lecture theatre of the Royal Institution. See Gooding (1985b), especially pp. 124–32.
2. On mapping inside and out of the laboratory, see Gooding (1989a); and on the testing of Pasteur's vaccine, Latour (1983).
3. An early example is Hobbes's criticism of the experimental method advocated by Boyle, discussed in Shapin and Schaffer (1985). By the nineteenth century the issues were somewhat different: see Galison and Assmus (1989a).
4. See Berman (1978) and Bence Jones (1862).
5. For examples in which new theoretical developments depended upon new technologies of observation, see Gooding (1989a) and Galison and Assmus (1989).
6. This is argued in Nickles (1989); see also Nickles (1985).
7. For examples, see the studies collected in Collins (1981).
8. On practice-ladenness, see Gooding (1989b).
9. 'Privacy' here connotes the state of experimentation prior to publication, rather than the activity of isolated individuals.
10. See Medawar (1964); and for the rhetorical aspect of accounts of experiment, Cantor (1989) and Naylor (1989).
11. Polanyi (1964). On the importance of procedures to thought, see Price (1980), Gooding (1982), Hacking (1983), Harrison (1978) and Tweney (1985); and on the communication of tacit knowledge, see Collins (1974). Jenkins (1987) and Ferguson (1977) deal with non-verbal aspects of discovery and invention.
12. On Lavoisier, see Holmes (1985, 1987); and for what Newton failed to explain in published accounts of his prism experiments, see Schaffer (1989).
13. Two contemporary examples are Goodfield (1982) and Latour and Woolgar (1979).
14. This is argued in Nickles (1984).
15. See Hackmann (1985), especially p. 105; and for an example of the interest in the constraining role of instruments, see Finn (1971).
16. For a detailed study of small-scale experiments, see Gooding (1989b); and for large-scale experimentation, see Pickering (1988).
17. Belloni (1970).
18. Hackmann (1978, 1979).
19. Settle (1961, 1983), especially pp. 12–14.

20. By 'recovering experiment' I mean the repetition of an experiment, not the attempt to produce an authentic replica.
21. The identity of an experiment is only fixed when it reaches the pages of textbooks (or what Ludwig Fleck called *vademecum* science: Fleck, 1979). The changing identity of experimental discoveries is discussed in Schaffer (1986) and Cantor (1989).
22. *Athenaeum*, 8 November 1845, p. 1080. For a more detailed account, see Gooding (1985a).
23. Herschel's letter to Faraday is reprinted in Williams *et al.* (1971), Vol. 1, pp. 461–3; his letter to Pepys (also of 9 November) is in the Royal Institution, Pepys Papers.
24. W. H. Pepys to J. W. F. Herschel, 12 November 1845, Royal Society (Herschel Papers). Pepys's battery is described in Pepys (1823).
25. For the importance of demonstration, see Hays (1983). Other aspects of the context of these experiments are discussed in Gooding (1989a).
26. Herschel (1830), pp. 76–7, 27.
27. For models of the dynamics of experimentation, see Pickering (1989) and Gooding (1989b). The Faraday example below suggests that this learning process is neither simple prediction nor unguided trial and error.
28. Faraday varied his experiments, often through sheer persistence rather than precise expectations about which parameters to vary. As Gassiot wrote to remind him, Faraday had urged others to try these experiments 'in every shape', with every possible variation: J. P. Gassiot to Faraday, 26 December 1845, reprinted in Williams *et al.* (1971), Vol. 1, pp. 478–9. Like Herschel, Gassiot had not taken the advice. Faraday had made this optical glass for the Royal Society two decades earlier.
29. J. W. F. Herschel to R. Hunt, 5 January 1846, Royal Society (Herschel Papers). See also letters between Herschel and Faraday of 15 and 22 January 1846, reprinted in Williams *et al.* (1971), Vol. 1, pp. 483, 486.
30. Shapin and Schaffer (1985).
31. Swenson (1972).
32. My earlier discussion of the plasticity of experiment should make clear that I do not think we can recover 'the actual route' to a discovery from laboratory records. This is because an element of reconstruction is inherent in the process of recording (whether sketching or writing). The activity is reflexive even when the experimenter is not consciously reflective. (For example, the need to record physical processes and the behaviour that elicits them in words and pictures enters into the process of observation itself: Faraday provided clues for himself in the phrases 'looking from above down' and the reminder that 'The rod in the circle is merely put there to shew the front and back part'). See Gooding (1989b), Chapters 3 and 5. This construction of experience differs from reconstructions introduced when writing up experiments for publication, discussed in Nickles (1988), Geison and Secord (1988), and A. Pickering, 'Editing and epistemology: three accounts of the discovery of the weak neutral current', forthcoming.
33. These figures reproduce pages from one of Faraday's early notebooks, now at the Royal Institution. This was subsequently incorporated by Thomas Martin into Faraday's *Diary* (see Martin, 1932–6). Faraday had assisted Humphry Davy with his electromagnetic researches (see Davy, 1821a, b) and had completed a thorough study of experimental response to Oersted's discovery of electromagnetism for Richard Phillips, the editor of the *Annals of Philosophy*: see Faraday (1821, 1821–2).
34. Work on Faraday's experiments on electromagnetism and electromagnetic induction was carried out by Jane Leigh and Susan Perrett as a final-year physics project at Bristol University. See Jane E. Leigh and Susan Perrett, 'A reproduction of

Michael Faraday's experiments on electromagnetism', Physics B.Sc. Stage 3 Project Reports, Bristol University, May 1985, unpublished. Further studies of the rotation experiments were subsequently made at the Royal Institution.
35. See Gooding (1989b), Chapters 6 and 7.
36. The artistry that enables seeing in science is discussed in Galison and Assmus (1989) and Gooding (1989b), Chapters 2 and 3.
37. These were carried out at the Royal Institution in 1987.
38. See Williams (1983, 1985) and Gooding (1989b), Chapter 2.
39. For a description, see Gooding (1985b).
40. Faraday had tried for the rotation of the wire about its own axis on 4 September 1821. Ampère's description in Ampère (1822) suggests that what rotated in his experiment was actually a thick cylinder, not a wire.
41. See op. cit. [note 34].
42. For a discussion of how experiments on other phenomena during 1831 prepared Faraday for the induction experiment, see Tweney (1985), pp. 204–6.

5

The Spirit of Investigation: Physics at Harvard University, 1870–1910

Lawrence Aronovitch

The purpose of the present chapter is to describe briefly an example of one fundamentally important root of research and development: the creation and early development of its venue, the research laboratory. As Frank James notes in his introduction to this volume, the dictionary definition of the laboratory is misleading. The case of the Jefferson Physical Laboratory at Harvard University in the late nineteenth century is one example of why this is so. The building, home to the university's physics department, was conceived with a mixture of motives; the product was a facility at which pre-eminent physical research would have profound effects on the science and technology of the twentieth century.[1]

1 The Beginnings of Laboratory Science

The laboratory as the home of the advancement of physical science and technology is a nineteenth-century invention.[2] While there were earlier precursors of the laboratory in terms of its research and teaching functions, it was in France, Britain and Germany during the first decades of the nineteenth century that scientists began to institutionalize their work in a distinctly *laboratory* setting. By the 1870s, both Oxford and Cambridge had completed the construction of buildings devoted expressly to physics with the Clarendon and the Cavendish laboratories, respectively.[3]

The American environment during the nineteenth century was somewhat less favourable to the development of physics as a discipline.[4] Much of this has been attributed to the emphasis in the new nation on applications and empiricism. Morse's telegraph (1837) and Singer's sewing machine (1851) seemed better representatives of the age than the eminent physicist Joseph Henry, who lamented in 1846 that 'our newspapers are filled with puffs of quackery and every man who can exhibit a few experiments to a class of young ladies is called a man of science.'[5]

Physics in the United States had not yet crystallized into a profession.[6] There existed no national physics society, nor was there a national physics journal, with which a community of colleagues could be cultivated.[7] In this respect, physics trailed behind other sciences such as chemistry, botany or geology. One reason for this was the state of physical knowledge in the late nineteenth century. The grand discoveries lay either in the past (Maxwell's laws) or in the future (relativity and quantum mechanics); the present offered a more mundane plateau. Meanwhile, the country's educational system was unable to address the needs generated by the era's advances in science and technology, a circumstance which the nation's educators had come to recognize by mid-century.[8] From this recognition came an early strand of the reasoning that led eventually to the foundation of the Jefferson Physical Laboratory.

2 Charles W. Eliot and John Trowbridge

In large part the driving force behind the establishment and evolution of the Jefferson Laboratory flowed from the relationship between two men: Charles William Eliot, the president of Harvard from 1869 to 1909, and John Trowbridge, a physics professor who served as the laboratory's director from 1888 to 1910. The work of these two men shaped the boundaries within which the nature of the laboratory could be established.

Eliot was a chemist who taught at the Harvard Lawrence Scientific School and studied at the new Massachusetts Institute of Technology in the 1860s. He had occasion to give the subject of higher education considerable thought, and shortly before becoming president Eliot published a widely read article entitled 'The New Education'.[9] In this article Eliot wondered what sort of practical education might be available to the current generation of young American men. He found that there were three strains of 'practical' education that incorporated the growing volume of scientific and technical knowledge of the day: schools such as Harvard's Lawrence Scientific School, established in 1847; the scientific courses of study within the established liberal arts colleges; and the new independent schools such as MIT, which opened its doors to students in 1865.

Institutions such as the Lawrence School Eliot condemned as inadequate for the needs of its students; he called them the ugly ducklings, principally because in their experimental states they were unable to compete with the established curricula of their parent colleges.[10] Furthermore, at these schools,

[i]t is quite possible for a young man to become a Bachelor of Science without a sound knowledge of any language, not even his own, and without any knowledge at all of philosophy, history, political science, or of any natural or physical

science, except the single one to which he has devoted two or three years at the most.[11]

Worse still were attempts to introduce a scientific 'course' within the college itself. At their core, the guiding philosophies of the liberal arts college and the polytechnic school were incompatible; the one stressed breadth of culture as well as learning and research for their own sakes, while the other emphasized learning and research with such specific goals in mind as the mastery of electrical engineering. To mix the scholarly spirit with the practical would spoil both.[12]

By far the best form of institution, in Eliot's view, was exemplified by MIT, where the members of the faculty were free to devote themselves entirely to the task of a practical education. The technical institute was the logical home of the practitioners of science, since here they could best seek out applications. At the same time, liberal arts colleges such as Harvard were more properly institutions devoted to the training of the mind, as had been expressed in an influential report of the Yale College faculty in 1829. The report stated that 'the two great points to be gained in intellectual culture, are the *discipline* and the *furniture* of the mind; expanding its powers, and storing it with knowledge. The former of these is, perhaps, the more important of the two.'[13]

The Yale report stressed the need for memorization and recitation of textbook passages and tended to ignore contemporary culture, history or science. The inquisitive and healthily sceptical atmosphere associated with intellectual pursuits by later generations was absent. Eliot did not fully agree with these forty-year-old prescriptions. The purpose of a liberal arts college, in his view, was to educate, to teach men how to think. This might be done partly with theology and Greek — but partly with chemistry and German as well. Upon becoming president of Harvard, Eliot expressed his intention of making the teaching of science an integral part of the college curriculum. For the college was obliged to teach science as well as the classics — although, in Eliot's opinion, the college had an obligation *only* to teach it: '. . . the prime business of American professors in this generation must be regular and assiduous class teaching'.[14] Indeed, any professor hoping to conduct research in addition to carrying out his teaching duties would, in Eliot's view, require 'fanatical zeal'.[15]

At the time of Eliot's inaugural in 1869, physics occupied a rather small position in the college curriculum, although other sciences were well represented. There was only one professor of physics, Joseph Lovering, who had held the venerable Hollis Chair of Mathematics and Natural Philosophy since 1838 and who shared the services of an assistant with Josiah Cooke, a chemistry professor. Lovering's method of teaching was a fine illustration of the preferences indicated by the Yale report: memorization and recitation with lectures and demonstrations. There was nothing equivalent, however, to the laboratory instruction offered by Cooke in chemistry.[16] Although he had engaged in some research, Lovering was dedicated to the teaching of physics as part of the classical curriculum.[17] Indeed, a later Harvard physicist, Edwin H. Hall, suggested that Lovering 'felt no more called upon to extend the domain of physics than as a preacher he would have felt obliged to add a chapter to the Bible'.[18]

Eliot's inaugural address suggested strongly that Lovering's teaching, while it enjoyed a good reputation, was by now outmoded. As Eliot remarked:

> The University recognizes the natural and physical sciences as indispensable branches of education, and has long acted upon this opinion, but it would have science taught in a rational way, objects and instruments in hand, — not from books merely, not through the memory chiefly, but by the seeing eye and the informing fingers.[19]

Eliot's reforms were implemented immediately, especially with the replacement of the prescribed course of study with the elective curriculum for which Eliot is perhaps best known. To facilitate a wide range of elective courses, an expanded faculty was required.[20] With respect to research, however, Eliot had nothing to say about its possible role in a college curriculum.

John Trowbridge joined the Harvard faculty in 1870 as an assistant professor of physics. Born into an old Massachusetts family in 1843, Trowbridge was one of the many new teachers brought to Harvard by Eliot during his first years as president and knew Eliot at MIT. Edward Pickering, a professor of astronomy, who also came to Harvard at the new president's invitation, was an important influence on the young physicist.[21]

More than any other person, John Trowbridge was responsible for the transformation of Harvard's physics department into a vehicle of laboratory teaching and original research. When informed that a certain textbook was out of date, Lovering had once remarked, 'That is just why I like it. Bringing it up to date means putting in a lot of improper matter.'[22] Trowbridge, on the other hand, was soon demonstrating his disposition for experiment. In 1871 he published an account of a new galvanometer of his own design which incorporated an improved method of measuring current intensities. In the following year he produced a short paper on the subject of animal electricity.[23] Trowbridge also introduced a course in experimental physics to Harvard. The final examination for the course consisted of ten questions, of which five required the collection of data. Although the questions were hardly revolutionary, they represented a significant break from the questions usually asked in lecture courses.[24]

Trowbridge shared Eliot's opinion concerning the need for radical changes in the teaching of physics, and this undoubtedly contributed to his appointment in the physics department. But his thinking carried him farther than Eliot had been inclined to go with respect to the function of the college laboratory. Whereas Eliot sought only a 'new' and modern education for his students which would teach them to *think*, Trowbridge was interested from the start in original research as well as the teaching of laboratory techniques for education's sake. As he moved forward in his Harvard career, Trowbridge found his opinions strengthening in this vein. But the existent facilities at Harvard were inadequate for anything more than an incidental pursuit of original research. Thus, Trowbridge began to think of the advantages that an endowed building devoted exclusively to physics would offer. He lacked only the trigger to give impetus to his project.

A laboratory for coursework was already located in Harvard Hall, where a physics lecture hall, a recitation room and an apparatus room for Lovering's instrument collection had also been established.[25] The laboratory facilities, consisting of 'working-tables, gas and water fixtures, and apparatus', were directed precisely towards the 'new education' sort of teaching prescribed by Eliot. But it was not a university research laboratory, and the modest number of research papers emanating from Harvard at the time were incidental to the principal business of teaching.[26]

By the early 1870s the physics department offered seven courses, of which three were requirements and four were elective. As the course selection grew, so did the staff. The first important addition after Trowbridge was Benjamin Osgood Peirce, who was hired in 1876. As Trowbridge's first research assistant, Peirce was introduced early in his career to experimentation and had published several papers while still an undergraduate.[27] In 1877 Peirce left Cambridge for Leipzig and Berlin in order to continue his studies at the doctoral level: research laboratories did not exist in America. While Trowbridge's efforts engendered some research during the 1870s, Harvard's emphasis was unambiguously on teaching, and Trowbridge had nothing with which to sell Eliot the need for a physics building. This situation was, however, about to change, and Trowbridge was to find a trigger.

3 The Need for a Physical Laboratory Building

In 1874 Daniel Coit Gilman became president of the new Johns Hopkins University in Baltimore. Gilman set out to build a university that would promote as many areas of research as possible, with instruction as thorough and as advanced as possible.[28] His first appointment to the new university's faculty was the physicist Henry Rowland. Rowland was one of the few physicists in the United States who was interested in 'pure science', as opposed to teaching or finding commercial applications of physical principles. In 1883 he defended his philosophy quite forcefully. Calling American science 'a thing of the future, and not of the present or past', he wondered 'what must be done to create a science of physics in this country, rather than to call telegraphs, electric lights, and such conveniences by the name of science'. but American education he found to be mediocre. He asked:

Shall our country be contented to stand by, while other countries lead in the race? Shall we always grovel in the dust, and pick up the crumbs which fall from the rich man's table, considering ourselves richer than he because we have more crumbs, while we forget that he has the cake, which is the source of all crumbs? Shall we be swine, to whom the corn and husks are of more value than the pearls? If I read aright the signs of the times, I think we shall not always be contented with our inferior position.[29]

Rowland was charged with helping to 'mature a plan for a physical laboratory and for the purchase and construction of instruments'[30] and with touring European university laboratories in order to see what might be applied to Johns Hopkins.[31] Rowland knew what he wanted, and that was to make research an integral part of a physics education:

Throughout the whole course [in advanced mathematical physics] an effort has been made to allow neither the experimental or the mathematical side of the subject to predominate but to so balance the two that the best knowledge of the subject may be obtained. In this respect it is believed that this course of instruction in Physics will meet a long-felt need in this country.[32]

A laboratory for these purposes, Rowland felt, must focus on apparatus and not on the building itself. He was contemptuous of the laboratory at Oxford, 'for as usual the architect had got the best of the physicist. There are some fine pieces of apparatus but as a whole they might far better have increased their stock than have spent as they did £10,000 for useless ornamentation.'[33] The balance of Rowland's European tour became a shopping expedition for apparatus that was meant to be *used*. As Rowland wrote to Pickering, 'they are nearly all for investigation and none of them are for amusing children'.[34] Henry's lament of 1846 had been heard.

One measure of Rowland's efforts is a brief list compiled by Trowbridge and some Harvard colleagues. Of the list's 192 items available for research use in the United States, 81 (42 per cent) belonged to Johns Hopkins and 69 (36 per cent) were specifically under Rowland's care in the physics department.[35] Rowland's collection whetted Trowbridge's appetite. Writing to President Gilman, he explained:

The collection of instruments of precision which the Johns Hopkins University possesses is far superior in all respects to that owned by any American university and indeed, in the subject of Electricity, larger and better than any university in Europe has in its possession. I am endeavoring to raise an endowment for a Laboratory of research in Cambridge and I shall use your list as a campaign document: for it makes our collection of apparatus look meagre indeed.[36]

Here was the needed trigger. For some years Trowbridge had been thinking of having a physical laboratory endowed at Harvard. In 1877 he published a brief pamphlet which stated his case: 'The "New Physics" of the present day requires a building and instruments which shall stand in the same relation to Physics that the equipment of an astronomical laboratory stands to astronomy.'[37] Harvard already boasted an observatory, a herbarium and a Museum of Comparative Zoology; Trowbridge felt that the university needed a research laboratory for the 'New Physics' of light, heat and electricity as well.

Trowbridge suggested that university professors were too preoccupied with teaching and should be free to pursue original investigations. Without the facilities devoted specifically to research which a physical laboratory could offer, however, few professors were likely to produce very much. Trowbridge sought to remedy the situation with a new building to house both teaching and research, recognizing that neither alone would do:

The department of Physics in a University must embrace both teaching and investigation. If it is given up entirely to teaching, the cause of science suffers, and the object of a University which is founded both to teach and to increase the sum of human knowledge is defeated. If it is given up to investigation entirely, science also suffers as well as teaching; for new minds are not educated to take the place of those who pass away.[38]

The call for a new laboratory evidently met with the university administration's approval, at least to a limited degree. Eliot agreed in his next annual report that Harvard needed a new laboratory, but he did not mention the need for research facilities as such:

The physical department has a strong claim for laboratories, cabinets, and lecture-rooms expressly designed for its use. In consequence of the makeshifts to which the College has been obliged to resort for several years past in the lack of lecture and recitation rooms, the department of physics is now divided between two buildings, neither of which is well adapted to its peculiar wants. This division is a serious hindrance to efficiency. Moreover, a modern physical laboratory requires so many special considerations in the building itself, that a good one can hardly be made in an old building not designed for such uses.[39]

Now that the President had agreed officially that a physical laboratory was desirable, Trowbridge had to find money. This search would occupy much of Trowbridge's time in the next few years. By 1881, Eliot was prepared to make the following announcement:

Last spring it was made known to the Corporation through Mr Alexander Agassiz that a friend of the university stood ready to build a physical laboratory at a cost of $115,000, provided that a permanent fund of $75,000 were raised, the income of which should be appropriated to the running expenses of the laboratory.[40]

The donor of the $115 000 was later revealed to be Thomas Jefferson Coolidge, a wealthy Boston businessman with interests in the New England textile industry and an alumnus who had graduated in 1850. Money for the current expense fund was to be collected over the next three years, a large part of it being donated by Agassiz, who was in charge of preparing plans for the building. Construction was completed in 1884 and the building was named the Jefferson Physical Laboratory in honour of Coolidge's great-grandfather, Thomas Jefferson.[41]

The Jefferson Physical Laboratory was uniquely designed with both research and teaching in mind.[42] It consists of three storeys as well as a basement. There are two wings, each some 20 metres square, connected by a narrower section about 14 metres wide and 22 metres long. The building lies north of Harvard Yard, its front facing south (see Figure 5.1). At the time the laboratory was built, the nearest street was over 90 metres away; it was hoped that vibrations from outside sources such as street traffic would thus be minimized. The Laboratory was deliberately plain, built of brick both within and without.

Each of the two wings of the Laboratory provided a locus for one of its two functions — instruction and research (see Figures 5.2–5.4). The east wing was devoted to the former. The laboratory for elementary instruction, which occupied the entire wing, was located on the third floor. Directly below was a large lecture room which seated 300. Students entered these rooms by means of a separate staircase on the east end of the building. The official reason for the staircase was 'to protect the rest of the building from the disturbances incidental to the movement of large numbers of persons',[43] but Theodore

Figure 5.1 View of the Jefferson Physical Laboratory from the south at the turn of the century. From Harvard University Archives.

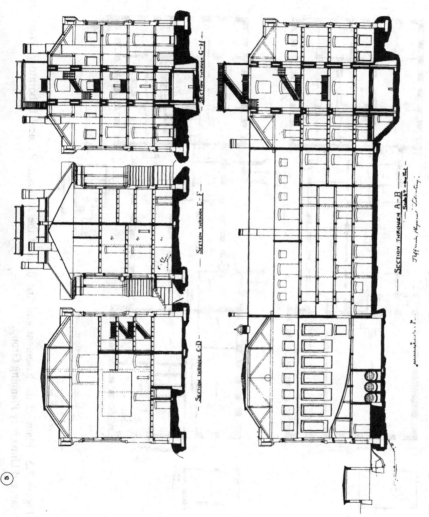

Figure 5.2 Cross-sections of the Jefferson Physical Laboratory. From the Harvard University Planning Group.

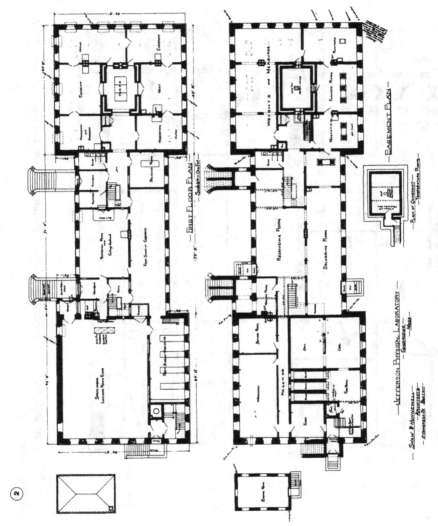

Figure 5.3 Plans of the basement and first floor of the Jefferson Physical Laboratory. From the Harvard University Planning Group.

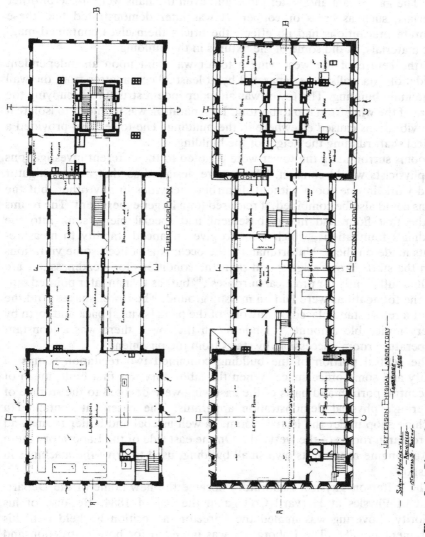

Figure 5.4 Plans of the second and third floors of the Jefferson Physical Laboratory. From the Harvard University Planning Group.

Lyman, a Harvard physicist of a later generation, suggested that 'the idea [was] that the undergraduates should be excluded from the main part of the building as much as possible'.[44]

The west wing was devoted to research. At considerable expense, no iron was used in its construction, so that magnetic disturbances might be minimized. The gas, steam and water pipes and even the nails were made of other materials, such as brass or copper. It was later demonstrated that these elaborate precautions had no effect; the bricks themselves contained magnetic materials, as did some of the fixtures in the building.[45]

In the centre of the west wing a tower was built upon an independent foundation. Its walls were separated by at least 30 centimetres from the wall of the main building. The tower was an independent structure occupying the centre of the wing from top to bottom. Its separation was intended to isolate it from vibrations caused elsewhere in the building. The tower also provided a vertical shaft running the height of the building.

Rooms surrounding the tower were devoted to the different investigations the physicists would conduct. These were designed so that an investigator could work in one room without disturbing activities in any other, but the rooms could also be combined, if required for a larger experiment. The rooms on the first floor and in the basement had special piers, sunk into the building's foundations, which were to give enhanced stability to measurements made on them. Unfortunately, the occupants noticed some vibrations from the start. Trowbridge stated that 'the conditions for steadiness . . . are fulfilled sufficiently for practical purposes',[46] but, as Lyman later pointed out, '. . . the foundations were laid on marshy ground. The tower shakes; and the top of a retort stand placed on any one of the piers [swings] back and forth by a very noticeable amount.'[47] Underneath the tower there was a 'constant temperature room', accessible by stairs from the basement.

The central portion of the building contained two recitation rooms, a library and some office space. When the laboratory was first built, much of the central portion and parts of the east wing were devoted to the storage of Lovering's physical demonstration apparatus. The basement contained a machine shop under the lecture room, as well as coal and boiler rooms and the research rooms of the west side. On the east side of the Laboratory there was an engine room in its own small building, used to drive the machines in the shop.

The Jefferson Physical Laboratory became the new home of the Department of Physics at Harvard College in the fall of 1884. Because of his seniority, Lovering was named the Director, a position he held until his retirement in 1888. The Laboratory was intended for both instruction and research, although these two functions were to be carried out independently of one another in the two wings of the building. On the whole, the Laboratory was well thought out in terms of 'its arrangement of space for scientific investigation, and its plant for the construction of new apparatus to meet the demands of the future'.[48]

Once the department moved into the new building, with its improved facilities, more courses came to require laboratory work. In particular, in the spring of 1886 the four departments of Chemistry, Botany, Zoology and Physics each established laboratory courses for incoming students. The purpose of these courses was 'to teach a science of observation by experimental methods to young persons whose mental training has been received almost exclusively through book-study of languages and mathematics'.[49] The physics course involved 'Measurements in Mechanics, Sound, Heat, Light, Electricity and Magnetism'.[50] At about the same time, the department offered graduate reading courses in physics for the first time.[51]

The elementary courses in experimental physics provided undergraduates with a sound introduction to the field. Secondary schools of the time, if they taught physics at all, gave scant attention to laboratory instruction. As a remedy, Edwin Hall compiled a 'Harvard Descriptive List of Elementary Physical Experiments' in 1887, which was distributed to secondary schools throughout the United States.[52]

Until this time few textbooks gave any attention to experiments. Trowbridge set out to correct this deficiency, writing:

> The experience of those who have made a careful study of the methods of instruction in physical science in the secondary schools, and who have judged of the results of these methods by the intellectual training manifested by students who present themselves for college, is unfavorable to the lecture or recitation system unsupported by laboratory work. . . . It is believed that the study of physical science, in a laboratory, affords an opportunity for this strenuous intellectual effort. The literary habit of mind is acquired by long study of language; and the scientific habit or instinct does not require less cultivation than the literary instinct.[53]

In terms of the improvements he was trying to fashion in physics education at the college level as well, these comments are a concise statement of Trowbridge's philosophy.

The development of Hall's list may well have been the most significant accomplishment at the Jefferson Physical Laboratory during Lovering's tenure as Director, for the list went on to recast entirely the method of teaching physics in American secondary schools. Between 1887 and 1912 seven editions of the pamphlet were published as it set standards for physics education.[54] A more effective secondary school course in physics could only lead to more advanced physics work in college and graduate school. Hall's pamphlet would lead to a new generation of students better equipped for physical research.

By the time Lovering retired, the physics department listed thirteen courses, all but two of which required at least two hours of laboratory work per week. The exceptions were Lovering's lecture course for new students and Trowbridge's course in mathematical physics.[55] Meanwhile, enrolment had jumped in ten years from a small handful to over 100 students.[56]

A moderate amount of research was accomplished while Lovering was Director of the Laboratory, but these years were devoted more to the improvement of the physics curriculum and the fulfilment of President Eliot's directives on science education. A new generation was being trained which would soon be able to take advantage of the opportunities offered by a building constructed with research explicitly in mind. The Jefferson Physical Laboratory represented a new stage in the gradual development of American institutions devoted to the systematic investigation of nature. In the next decades notable work would be done. More importantly, the work would contribute to the evolution of the American science community. The university research laboratory was firmly establishing itself.

4 Encouraging a Spirit of Investigation

With the exception of Trowbridge's own efforts, little physical research was conducted at Harvard University before the Jefferson Physical Laboratory was built. When Eliot invited Edwin Hall to join the Harvard faculty in 1881, he wrote that the job would be a fine opportunity 'for a person who can lecture and also give laboratory instruction', but he did not mention research opportunities at all.[57] In a later letter, Eliot mentioned almost as an afterthought the hope 'that your activity in physical research would not entirely cease', but certainly Hall's attention was expected to focus on instruction first.[58]

After the first year of the Laboratory's operation, Lovering made the following report on research:

These investigations were in the main upon subjects which can be studied from year to year: such as the Moduli of Elasticity; Electromotive Force; the Electrical Resistance of Electrolytes; the Mechanical Equivalent of Heat; Quantitative Measurements in Magnetism; and the Relation of Light to Electricity. By this method [Professor Trowbridge] hoped to make the laboratory a centre of systematic work in Physical Science.[59]

Trowbridge succeeded Lovering as director of the Laboratory in 1888. From the start of his tenure Trowbridge made his influence felt. Every course required laboratory work, with the exception of the introductory course and the advanced course in electrodynamics and electromagnetism. The first of these would be dropped from the curriculum in a year and the second was sufficiently advanced for it to be expected that any student would have some laboratory experience if he was to be prepared for the course. A decade earlier the only physics course requiring laboratory work had been a novelty; now it was the standard. In his first report to President Eliot, Trowbridge indicated that he intended to emphasize both improved laboratory instruction and better research. He remarked with satisfaction that 'the Jefferson Physical Laboratory is not excelled by any laboratory in Europe in respect to its arrangement of rooms, its isolated position, its piers, and its elementary laboratories'.[60]

Money was a major concern for Trowbridge, as it is for all laboratory directors. He pointed out that the Laboratory's income barely met the building's basic running expenses[61] and declared:

It is true that occasionally a genius can discover great laws by the use of bits of string and magnets, but the systematic work of a physical laboratory, like that of an astronomical observatory, requires a large endowment.[62]

Shortly thereafter a new fund was established with a principal of $7720 contributed by some fourteen donors, including corporate sponsors such as the American Bell Telephone Company ($1000) and well-to-do private benefactors such as Theodore Lyman ($100). The fund, named for Lovering, was to provide income to be used specifically 'for the promotion of original research at the Jefferson Physical Laboratory' — an ironic stipulation in the light of Lovering's unenthusiastic attitudes towards such activity.[63]

The Lovering fund was too modest to provide more than a few hundred dollars a year.[64] Thomas Jefferson Coolidge remedied the problem in 1901, when he provided an additional $57 500 to be used 'primarily for the laboratory expenses of original investigations by members of the Jefferson Physical Laboratory staff'.[65] No further endowments were established while Trowbridge was Director.

The need for more money devoted specifically to research reflects the new direction the Laboratory took under Trowbridge. In the 1870s Lovering had spent about $1500 a year on physical apparatus purchased in Europe, which he then stored in cabinets until he required them for lecture demonstrations. In contrast, Trowbridge spent less than $100 a year on such orders when he became Director. Instead, he expanded the machine shops in the basement of the Laboratory. If a professor required a certain instrument in his laboratory, whether for his own research or for a course he was teaching, he had available the means to construct a prototype in the shop and improve its design until it met his satisfaction.[66]

While Trowbridge was Director, research became the central and predominant function of the Laboratory, as illustrated by the fact that the Lovering and Coolidge funds were directed expressly towards research. To Trowbridge, 'the life of a physical laboratory consists in its spirit of investigation'.[67] Before Eliot's presidency, research was not at all the rule at Harvard, but in the early years of the new century Trowbridge could assert with pride that 'the Laboratory has now the best equipment of any Physical Laboratory either in this country or abroad' for purposes of research.[68] Most of the research output of the time was not especially memorable, although there were significant highlights. The importance of the Laboratory at this time lay more in its presence as an opportunity for a wide range of physical investigations in the United States and as a training ground for future generations of physicists.

Trowbridge himself produced an impressive number of research papers, in spite of heavy teaching and administrative duties.[69] His interests were drawn to spectral analysis as well as X-rays, which Röntgen had discovered in 1895. One Trowbridge paper of 1897 on the subject of X-rays is of interest less for its subject matter than for a piece of equipment built for the paper's experiments.[70] At the beginning of the paper Trowbridge observed that 'the value of a large storage battery for the study of the discharge of electricity through gases has long been recognized'. He then described in detail the construction of a storage battery at the Jefferson Physical Laboratory consisting of 10 000 cells, far larger than any other battery of the time. For many years thereafter, the battery was a famous and unique Harvard fixture, attracting researchers from far and wide.

More important than Trowbridge's own research was his ability to foster research in others. According to Hall, he had 'keen perception, he saw the capabilities of other men, and was glad to give them opportunity'.[71] Lyman said that Trowbridge's great gift was 'a sort of scientific clairvoyance' coupled with the ability to inspire students.[72] Eighteen graduate students received Ph.D. degrees while Trowbridge was Director, of whom four were Laboratory staff members when Trowbridge retired.[73]

Benjamin Peirce had been Trowbridge's first research assistant in the 1870s. When he returned to Harvard from his studies abroad, his time was divided between the physics and mathematics departments. In addition to his work in mathematics, including his well-known *Short Table of Integrals*, Peirce engaged in two general areas of physical research, the conductivity of poor conductors (such as stone) and magnetism.[74]

The next most senior member of the faculty was Edwin Hall. When Hall decided in 1877 to become a physicist, he thought of studying at Harvard, but Trowbridge discouraged him by arguing that the facilities there were inadequate. Instead, Hall went to Johns Hopkins, where as a graduate student he discovered the effect (which now bears his name) of a magnetic field on electric currents in conductors.[75]

At Harvard, Hall invested much of his time in work other than research, notably physics education.[76] During Trowbridge's tenure as Laboratory Director, Hall did not conduct any further research into the Hall effect, except for a paper of 1893; instead, he pursued research in thermal phenomena, including the thermal conductivity of metals (in contrast to Peirce's work with non-metals), the thermodynamic behaviour of liquids and a variety of thermoelectric effects.[77] One inquiry of Hall's illustrates well the forethought with which the Laboratory was designed. Hall sought to answer the very old question, 'Do falling bodies move south?' His investigation made use of the independent tower in the west wing of the building, in which he dropped balls from a height of 32 metres.[78]

Perhaps the most intriguing career of Trowbridge's tenure was that of Wallace Sabine. In 1895 the university had erected a new museum building in Harvard Yard. Unfortunately, the lecture room of the museum possessed lamentably bad acoustics, and President Eliot asked Sabine to do what he could to improve the situation. Sabine's brilliant work, carried out in the lecture room itself as well as in the Laboratory's lecture hall and constant-temperature room, so impressed Eliot that the President recommended Sabine as a consultant to the builders of Boston's new Symphony Hall.[79] In Edwin Hall's opinion: 'Sabine's development of architectural acoustics from a condition of gross and ineffectual empiricism to the status of a reasoned and fairly exact science is probably the most notable single achievement in the history of the Jefferson Laboratory during its first three decades.'[80]

One fruitful line of research was Theodore Lyman's work on hydrogen spectra. This work, which began in preparation for his doctoral dissertation of 1900, led in 1914 to the discovery of the Lyman series of spectral lines for atomic hydrogen, which provided experimental evidence for Niels Bohr's hypothesis of a quantized atomic structure. The Lyman lines were seen to represent the electron transitions to the $n = 1$ state.[81]

Another notable physicist of the time was George Washington Pierce. His speciality was electromagnetic radiation and wireless telegraphy. His studies in piezoelectricity and magnetostriction led to useful radio applications which left Pierce, who held the patents, in very sound financial health.[82] Other investigators included H. N. Davis, who in 1907 won the university's Bowdoin Prize for an essay examining the motion of a violin string, and H. W. Morse, who was interested in voltaic cells and electrolysis.[83]

In 1908 Percy Bridgman received his Ph.D. for a paper entitled 'Mercury resistance as a pressure gauge'. Although he conducted most of his pioneering work in high-pressure physics after Trowbridge had retired, Bridgman started his investigations and received his training in the laboratory that Trowbridge built.[84] Trowbridge provided the facilities and opportunities which culminated in the award of the Nobel Prize to Bridgman in 1946, the first won by a member of the Harvard faculty in physics.

In 1903 the Laboratory began systematically to issue the research papers of its staff in an annual collection of reprints entitled *Contributions of the Jefferson Physical Laboratory*, which it distributed to physicists around the world. The early editions of the *Contributions* serve as a map for the growth of Harvard's physics department as a seat of research under Trowbridge's direction. Forty years earlier the physics faculty had consisted of Lovering and Trowbridge alone, with neither producing any original research. This was no longer the case, as Trowbridge explained in 1896: 'To-day a certain amount of original work is expected of [the physics professor at Harvard]. During the past ten years more original research in Physics has been done in Harvard University than in the previous two hundred years.'[85] By 1910 the quality of research was judged unsurpassed in America, and the Jefferson

Physical Laboratory was well on its way to becoming one of the most respected centres of physical research in the world.[86]

5 Conclusion

The building of the Jefferson Physical Laboratory marked the beginning of a new period for physics in the United States. Charles Eliot was primarily interested in the promotion of laboratory instruction as a part of the 'new education' he brought to Harvard in 1869. An expanded curriculum required the larger and more appropriate quarters afforded by the Jefferson Laboratory. To John Trowbridge, this was not function enough. In addition to improved instruction, Harvard's department of physics required active researchers if it was to be a leader in physics in the United States.

In one sense the work being done reflected the general state of American physics in the first years of the new century, in that it was experimental and not theoretical in nature. Americans were predominantly measurers. Michelson and Morley had measured the velocity of light, Lyman was measuring the spectral lines of hydrogen at Harvard, and soon Millikan would be measuring the charge of the electron. Sabine's acoustical work was purely experimental, as was Bridgman's work in high-pressure physics.

The Jefferson Physical Laboratory under Trowbridge may have been the seat of experimental physics, but it also laid the foundations for a new generation of physicists as ready as any of their European colleagues to meet the challenging problems of twentieth-century physics, whether the questions were experimental or theoretical. In this sense, the significance of the Laboratory lay in the opportunities presented by a well-equipped facility in which a diversity of investigations was vigorously encouraged. The new generation of Harvard physicists included experimentalists such as Lyman and Bridgman, but it also encompassed men such as Edwin C. Kemble, the quantum theorist.

The Jefferson Physical Laboratory is a *place*. With its sister laboratories at the turn of the century, it offered a setting for physical research, and with it the institutionalization that the discipline needed to progress. At the end of the 1800s, the Jefferson Laboratory had the good fortune to be an early site for American research, largely by virtue of the work of John Trowbridge, and to lay the groundwork for the more mature research of the new century.

Notes and References

1. Some of the themes presented here are explored in greater detail in Aronovitch (1983). See also Holton (1984).
2. Phillips (1983).
3. Phillips (1983), p. 500. See also Mendelsohn (1964).
4. Shryock (1948).

5. Quoted in Kevles (1978), p. 4. See also Kohlstedt (1976), p. 2, and Storr (1953), p. 5.
6. Daniels (1967) identifies four stages in the process of professionalization: pre-emption, institutionalization, legitimation and attainment of professional autonomy.
7. Although the physics community in the United States did not yet enjoy all the attributes of professionalization, the same cannot be said of the American scientific community in general. Throughout much of the nineteenth century, scientific societies and journals were being founded and funds were sought for the furtherance of science. See Daniels (1968) and Reingold (1972).
8. A good discussion of this point may be found in Veysey (1965).
9. Eliot (1869).
10. Ibid., p. 206.
11. Ibid., p. 210. One might observe that similar complaints continue to be voiced today about education in the sciences.
12. Ibid., pp. 214, 215.
13. Quoted in Storr (1953), p. 2.
14. Charles William Eliot, Inaugural Address as President of Harvard College, reprinted in Pusey (1969), p. 21.
15. Quoted in Morison (1936), p. 378.
16. Annual Report of the President of Harvard University 1867–1868, p. 29. For the course of study in chemistry, see p. 28.
17. Details of Lovering's research may be found in Peirce (1909), pp. 335–7.
18. Hall (1930), p. 277.
19. Eliot, op. cit. [note 14], p. 4.
20. The growth of Harvard's teaching staff upon Eliot's inauguration is documented in the Annual Report of the President of Harvard University 1871–1872, p. 10.
21. Hall (1932), pp. 185–7.
22. Hall (1930), p. 278.
23. Trowbridge (1872).
24. The course is listed in the Harvard University Catalogue 1872–1873, p. 60. The final examination appears on p. 282. Students were required to measure resistance, use a galvanometer, explain the principle of the Wheatstone Bridge, and so forth.
25. See Annual Report of the President of Harvard University 1869–1870, p. 25.
26. Trowbridge reported that the physics department produced 34 papers (of which 14 were his own) from 1871 through 1876. See Trowbridge (1877?), pp. 5–7.
27. Hall (1919), pp. 444–5.
28. Gilman (1906), p. 41. See also French (1946) and Hawkins (1960).
29. Rowland (1883).
30. Gilman to Rowland, 21 June 1875. Quoted in Miller (1970), p. 91.
31. Gilman (1906), pp. 14–15.
32. Quoted in Miller (1970), p. 196.
33. Rowland to Gilman, 14 August 1875 (Daniel Coit Gilman Papers, Milton S. Eisenhower Library, Johns Hopkins University; hereafter DCG).
34. Rowland to Pickering, 17 May 1876. Quoted in Hawkins (1960), p. 46.
35. Gibbs et al. (1879). Together, Harvard and Johns Hopkins dominate the list, with 139 items. The list makes no claim to be comprehensive. The authors' letter to colleagues asking for contributions to the list, which is published as well, refers only to 'such apparatus as you possess . . . and which you are willing to place . . . at the disposal of any properly qualified person for the purposes of investigation'. What respondents chose to omit is not known.
36. Trowbridge to Gilman, 11 August 1879 (DCG).
37. Trowbridge (1877?), p. 1.

38. Trowbridge (1877?), p. 4.
39. *Annual Report of the President of Harvard University 1878–1879*, pp. 46–7.
40. *Annual Report of the President of Harvard University 1880–1881*, p. 38. The Corporation is the governing body of Harvard University.
41. *Annual Report of the President of Harvard University 1883–1884*, p. 43. In addition to his political activities, Thomas Jefferson was interested in the sciences and served as a president of the American Philosophical Society.
42. The description is based for the most part on Trowbridge (1885), as well as on the architectural plans and cross-sections of the building. Since it was built the Laboratory building has undergone considerable modifications and many of its original features no longer exist. See also [Lyman] (1932).
43. *Harvard University Catalogue 1884–1885*, p. 252.
44. Theodore Lyman, 'Recollections', mimeographed (Cambridge, Mass.), p. 2.
45. Willson (1890).
46. Trowbridge (1885), p. 231.
47. Lyman, *op. cit.* [Note 44], p. 1.
48. *A Guide to the Grounds and Buildings of Harvard University* (Cambridge, Mass., 1893), p. 85.
49. *Annual Report of the President of Harvard University 1885–1886*, p. 21.
50. *Harvard University Catalogue 1885–1886*, p. 103.
51. *Harvard University Catalogue 1883–1884*, pp. 94–5. The courses were 'Thermodynamics and Molecular Physics', offered by Wolcott Gibbs, and 'Theory of Sound', offered by Trowbridge.
52. Hall (1887); *Annual Report of the President of Harvard University 1886–1887*, p. 138.
53. Trowbridge (1884), pp. iii–iv.
54. Bridgman (1941), p. 78; Webster (1938); Hall (1930), p. 286.
55. *Annual Report of the President of Harvard University 1887–1888*, p. 69. Trowbridge's course focused on Maxwell's mathematical work in electricity and magnetism.
56. *Annual Report of the President of Harvard University 1887–1888*, p. 69.
57. Eliot to Hall, 2 March 1881 (Charles W. Eliot Papers, Harvard University Archives, Harvard University; hereafter CWE).
58. Eliot to Hall, 14 April 1881 (CWE).
59. *Annual Report of the President of Harvard University 1884–1885*, p. 156.
60. *Annual Report of the President of Harvard University 1888–1889*, p. 172.
61. *Annual Report of the President of Harvard University 1888–1889*, p. 171. According to the *Statement of the Treasurer of Harvard College 1889*, p. 42, the income of the laboratory for the year was $3920.29 from the endowment. There were also supplements including an annual payment from the college of $600.00 and the income from a small 'Jefferson Physical Laboratory' endowment which at the time (1 August 1888) had a principal of $1718.31.
62. *Annual Report of the President of Harvard University 1888–1889*, p. 172.
63. The Joseph Lovering Fund for Physical Research is first mentioned in the *Statement of the Treasurer of Harvard College 1891*, p. 24. See also *Endowment Funds of Harvard University* (Cambridge, Mass., 1948), p. 63.
64. In 1891–2, for example, when the fund had grown to $8168.09, its income was $405.77. See the *Statement of the Treasurer of Harvard College 1892*, pp. 27, 45.
65. *Statement of the Treasurer of Harvard College 1902*, p. 10. See also *Endowment Funds of Harvard University*, pp. 41–2, which cites further stipulations of the fund.
66. See Trowbridge's discussion in *Annual Report of the President of Harvard University 1890–1891*, pp. 169–171.
67. Trowbridge, in *Harvard Graduates Magazine*, **6** (March 1898), p. 394.

68. Trowbridge to Eliot, 3 March 1902 (CWE).
69. Between 1871 and 1911 Trowbridge produced 83 papers, many of which were written in collaboration with his students. See the bibliography in Hall (1932), pp. 201–4, and Thaddeus J. Trenn, 'John Trowbridge', *DSB*, **13**, 473.
70. Trowbridge (1897).
71. Hall (1930), p. 287.
72. Lyman, *op. cit.* [Note 44], p. 1.
73. *Harvard University Department of Physics Doctors of Philosophy and Doctors of Science, 1873–1964* (Cambridge, Mass.: Harvard Graduate Society for Advanced Study and Research, 1965) pp. 5–6.
74. Hall (1919), pp. 460–2.
75. Bridgman (1941), pp. 73–5, 81.
76. Already mentioned is Hall (1887). Also of interest are Hall and Bergen (1895) and Hall (1913). The text by Hall and Bergen is based largely on the 'Descriptive List'.
77. Bridgman (1941), pp. 79–83.
78. Hall (1903, 1904).
79. Daniel J. Kevles, 'Wallace Clement Ware Sabine', *DSB*, **12**, 54. See also Beranek (1977).
80. Hall (1930), p. 289.
81. Lyman (1914).
82. Charles Süsskind, 'George Washington Pierce', *DSB*, **10**, 605.
83. Stanley Goldberg, 'History of Physics at Harvard University', unpublished manuscript (Cambridge, Mass., 1962), p. 5.
84. Kemble and Birch (1970), pp. 27–35.
85. John Trowbridge, in *Harvard Graduates Magazine*, **4** (June 1896), 609.
86. Cattell (1910), p. 685. The ten strongest centres of American physics in Cattell's list are, in order: Harvard, the Bureau of Standards in Washington, Princeton, Johns Hopkins, the University of Chicago, Columbia, MIT, Cornell, the Carnegie Institute and the Department of Agriculture.

6

J. J. Thomson and 'Cavendish Physics'

Isobel Falconer

In this chapter I do not discuss the building of the Cavendish Laboratory. The original building, designed mainly by Maxwell, was fairly well documented and, although two extensions were built during the period I am talking about, they simply continued in the pattern Maxwell and Fawcett had set. No new thought or philosophy went into the Laboratory buildings until the construction of the Mond Laboratory and the Austin wing in the 1930s.[1]

Instead, I discuss a change in the identity, almost the definition, of the Laboratory around the turn of the century; a change in what the phrase 'The Cavendish Laboratory' meant to the outside world. This reflected changes within the Laboratory, rather than in its physical structure. The change can be summed up by noting that in the nineteenth century we talk about 'Cambridge' mathematical physics, but by the 1920s and 1930s of 'Cavendish' physics — two very different things.

The change took place during J. J. Thomson's long professorship and the origins of many of the later characteristics of 'Cavendish' physics can be seen in his own work. This chapter examines what, and how, Thomson contributed to the changing image of the Cavendish, and thus provides an analysis of the way social, psychological and scientific influences interacted to change the character of the Laboratory. First, I characterize the change by comparing the work of a physicist under Thomson's predecessor, Rayleigh, who was Professor from 1879 to 1884, with a physicist under Rutherford, Thomson's successor, in the 1920s and 1930s. Second, I look at Thomson's own approach to physics and what he contributed to the change. Third, I discuss how this change came about in terms of Thomson's influence and the teaching at the Laboratory.

1 The Change from C19 'Cambridge' Physics to C20 'Cavendish' Physics

The keynote of laboratory physics in the 'Cambridge' tradition was a close match between detailed mathematical analysis and precise experimental measurements. The norms were set out by William Thomson (Lord Kelvin) and Tait in their 'rules for the conduct of experiments':

> When a particular agent or cause is to be studied, experiments should be arranged in such a way as to lead if possible to results depending on it alone; or, if this cannot be done, they should be arranged so as to show differences produced by varying it.[2]

In a similar vein, Schuster and Shipley summarized the approach of the Cambridge school, which

> . . . clearly defines a problem confining it to such limits, wide or narrow, as will convert it into a precise problem which can be formulated and submitted to mathematical analysis. There must always be a definite answer to a definite question, and, unless the mathematical difficulties are insuperable, the consequences of any assumptions may be obtained in a form in which they can be tested, not only as to their general nature but also as to their numerical values. The result may not be far-reaching, but within its limited field it is definite.[3]

Thus, experiments must be well controlled, quantitative and precise, and their accuracy was analysed in some detail.[4]

The theories of the time attempted to reduce all phenomena to forms of matter in motion, the aether being the unifying and fundamental matter. They were often based on mechanical analogies whose consequences were worked out in considerable detail with some of the most advanced mathematics of the day.[5] The use of such analogies ensured a 'physical understanding' of what was going on. Physicists were not to be seduced by the elegance or logic of mathematics alone. Such theories had to meet the requirement of close agreement with precise quantitative experiment. Unsubstantiated theoretical leaps were not allowed (in principle).

Rayleigh, who held the Cavendish Professorship from 1879 to 1884, felt that the work of the Laboratory needed a focus which exemplified these norms and embarked on a programme of measurement of electrical standards. In practice, this particular programme failed to attract researchers and Rayleigh had to co-opt his sister-in-law to help him take measurements. However, most of the rest of the work of the Laboratory was in a similar tradition: precise measurement of physical constants.[6]

The apparatus with which the experiments were performed was frequently home-made. Yet this was largely because of the pioneering nature of the work and lack of commercially produced apparatus, rather than a deliberate policy of parsimony. The Laboratory did buy commercial apparatus when available — in particular, electrometers and spectroscopes. Special apparatus was engineered by skilled mechanics, either in the Laboratory's own work-shops (which Rayleigh extended) or in the University's engineering work-shops. Occasionally an instrument maker would be commissioned.

Thus, the successful physicist of the Rayleigh era had to be an accom-plished mathematician, and virtually all research workers had taken the Cambridge Mathematical Tripos. Yet he needed a grasp of the physical significance of his mathematical manipulations. He required a high apprecia-tion of experimental design, being aware of the capabilities and limitations of his apparatus, and had to be good with his hands in order to make delicate adjustments. Above all, he had to be patient and tenacious, for the experiments were finicking and difficult to perform, yet the results were very limited. Imagination was not much called for.

Compare this with the situation in the 1920s and 1930s, when the Laboratory was largely turned over to atomic and nuclear physics.[7] Tremendous excitement about physics, and extreme financial stringency, are the outstanding characteristics in students' reminiscences. Experimental 'facts' were still all-important in Rutherford's eyes, yet the nineteenth-century requirement for precision was gone. An order of magnitude was good enough for comparison with theory. Theoretical development was guided largely by 'intuition', backed up by fairly elementary mathematics and sometimes illustrated by analogy. But imagination and a 'feel' for what was going on were stressed. (Rutherford himself could 'almost see' electrons.[8]) The logic of mathematics provided subsequent justification. The phenomena were not always well understood and, hence, the problems not 'well defined' in Schuster's sense, but the results were nearly always 'far-reaching'. The experimental emphasis was on 'simple' apparatus which the student could construct for himself, and shortage of money necessitated using materials again and again. Vacuum and discharge techniques were the stock in trade of the experimenter. This was the era of 'string and sealing wax'.

Thus, in Rutherford's time a successful physicist needed to be numerate, but did not need to be an advanced mathematician. He required the moderately good, but not unrestrained, imagination which endowed him with physical intuition. He had to be a good glassblower and handyman to construct his apparatus, but precision engineering was not called for. Delicate adjustments to apparatus were less prominent, but patience was called for instead in tracing electrical connections and leaks in the vacuum systems. However, the rewards for patience were likely to be much greater in 'far-reaching' results. Table 6.1 summarizes the differences between the Rayleigh and Rutherford eras.

2 J. J. Thomson's Approach to Physics

Thomson's own experimental work marked the break with the 'Cambridge' tradition, despite his having been educated within this tradition. The hall-

Table 6.1   Summary of the main characteristics of experimental physics at the Cavendish Laboratory under Rayleigh and Rutherford. Asterisks indicate characteristics attributable to Thomson

Rayleigh	Rutherford
Standards	*Atomic physics
Precision	*Orders of magnitude
Advanced maths	*Elementary maths
Adequate money	*Stringency
Commercial apparatus	*Homemade apparatus
Electrometers, revolving coils, Wimshurst machines, spectroscopes	*Glassblowing, *vacuum techniques, counters, on-line power
Investigation of details	*Fundamental investigations
Perseverance and dedication	*Enthusiasm and excitement

marks of his work were extremely speculative theories, inspired by late-nineteenth-century 'Cambridge' concepts and analysed in great mathematical detail, yet of which he required only the broadest agreement with imprecise experiments. Indeed, he exhibited an almost complete inability to match theory and experiment. He was not interested in quantitative precision, or in a close match between theory and experiment, being satisfied with a rough comparison of orders of magnitude. His contemporaries admired him for his fertile imagination and he was noted for proposing far-reaching theories on the basis of the slenderest experimental evidence, or even no evidence at all. He did not worry particularly if these theories subsequently turned out to be wrong. His theories all originated in mechanical analogies, usually based on vortices in the aether, and he emphasized the value of such theories for giving a grasp of the general features of any phenomenon. Mathematical analysis came later, for elucidating details.[9]

I have identified several origins of Thomson's new, and idiosyncratic, approach to experiment. There were three important, unique factors in his personality.

First, he lacked both the patience and the manipulative ability to be a successful 'Cambridge' experimenter. His early attempts in this direction were failures: his measurement of emu/esu, for instance, was 1 per cent too low, and his biographer, the fourth Lord Rayleigh, criticized him for being 'over sanguine that he had foreseen the possible sources of error, without applying the test of using alternative methods'.[10]

Second, he was endowed with an extremely fertile imagination and a great enthusiasm for far-reaching theories, and he was not prepared to confine himself to a 'limited and definite range'.

Third, his imagination was strongly visual, going beyond the usual Victorian predilection for visualizable analogies and determining the type of experiments he did and the apparatus he used. He had a marked preference for visual, rather than metrical, results and would seldom sacrifice visibility for the sake of quantification. Thus, for instance, in his positive ray experiments he preferred to measure the discharge potential by observing the rate of sparking across a parallel spark-gap rather than use an electrometer, and to detect the rays with a photographic plate rather than use a Faraday cylinder, which gave a count of the actual number of rays collected.[11]

These three factors led Thomson to investigate electrical discharge through gases, a topic which, at that time, no serious academic would touch. As Schuster later recorded, it was considered fit only for 'cranks and visionaries',[12] probably because, as Schuster also judged in 1884, '. . . the qualitative phenomena have not . . . been sufficiently separated from the great number of disturbing effects to allow us to give a decisive value to quantitative measurements'.[13] Thus, definite problems appeared impossible and precise quantitative measurement seemed meaningless. Most results were obtained by observing the luminosity in the gas or fluorescence on a screen, methods ideally suited to Thomson's visual predilection. His theoretical work on the vortex atom indicated that electrical discharge might reveal the fundamental connection between electricity and matter, so he pursued it regardless of academic disapproval.[14]

This brings us to a fourth, very important, reason why Thomson diverged from the norms of the 'Cambridge' school. When he entered the field of discharge research, he became subject to a very different set of standards. The 'discharge research group' was largely composed of amateur scientists: De la Rue, Muller, Spottiswoode, Moulton and Crookes. (Though the latter was a professional, he was nevertheless not mathematical.) They had very little mathematical training and worked with largely qualitative theories. Quantitative experiments were rare, most results being recorded in the form of diagrams of the appearance of the discharge. Precision measurement was unheard of.[15] Thomson's approach to physics was thus largely the result of the interplay between the two very different sets of standards to which he was subject: those of the 'Cambridge' school and those of the 'discharge research group', combined with his personal characteristics, his impatience and desire for far-reaching results, his imagination and a strong preference for visual results.

Thomson also had to acquaint himself, and the Laboratory, with new experimental techniques, glass discharge tubes and vacuum pumps being the standard tools of the discharge experimenter. Such apparatus was virtually unknown in the Laboratory prior to Thomson's era and none of the technicians or staff had any expertise in glassblowing. Indeed, Thomson seems to have had to scour Cambridge to find anyone at all competent, and his own work was often held up for lack of a glassblower.[16]

Thus, in Thomson's experimental work we can identify the origin of a number of the characteristics of Rutherford's Cavendish: the use of vacuum and discharge techniques, the absence of precision experiment, comparison of orders of magnitude for matching experiment and theory, the investigation of the fundamental structure of matter, and an encouragement of theoretical speculation (although no one subsequently was quite as unrestrained as

Thomson) based on physical intutition.

However, Thomson's use of secondary mathematical analysis was largely lost, mainly owing to changing patterns in the training of British physicists. They generally took science for their first degree, rather than mathematics, and few had the skill and understanding to appreciate the more abstract and mathematical theories in physics. Similarly, they entirely discarded the commitments and concepts on which Thomson's theories were founded, few being able to appreciate a pre-electron physics based on the now unfashionable aether.[17]

Thomson also seems to have initiated the policy of parsimony in the Laboratory. Again, he invites comparison with his predecessor. Finding the Laboratory under-equipped, Rayleigh promptly set about fund-raising, seeking donations from his friends and acquaintances, and raising about £1500 for apparatus. During his five-year tenure he also added to the establishment one demonstrator (in 1880) at £100 per year, and two assistant demonstrators (in 1884) at £60 per year each, all paid for by the University. In contrast, Thomson waited 8 years before building a badly needed extension to the Laboratory in 1896, until he had scraped together the necessary £2000 by savings from laboratory and lecture fees. He had previously made a few enquiries which suggested that the University would not willingly put up the money, and he made no attempt to push the issue or attract money from elsewhere. Similarly, the extension of 1908 was not contemplated until Rayleigh offered the Cavendish £5000 from his Nobel Prize money. To this Thomson again added £2000 saved from fees. During the 35 years Thomson was Professor, the establishment increased only by two University-paid lecturers. The holders of both these positions held other stipendary posts in the University and the lectureships cost the University only £50 per year each. A number of extra demonstratorships were established and an ever-increasing number of technicians employed, but all these were paid for from fees.[18]

It is evident that Thomson was averse to going out and asking for money. Indeed, there is evidence that he even refused it when offered. A letter from Thomson to Oliver Lodge in 1905 explains his reasons for refusing an offer of Government money: 'my income is sufficient for my needs' and 'I have two private assistants and any number of pupils who would help me to work out anything that seemed promising so that money would not help me there'.[19] This letter highlights two further differences of outlook between Thomson (and later Rutherford) and Rayleigh.

First, Thomson regarded his personal income as his own, although he might spend some of it on furthering his own research. He never made donations to the Laboratory, even when he won the Nobel Prize. It did not occur to him that extra money from the Government could be used for the general good of the Laboratory. Rayleigh, on the other hand, had contributed £500 of his own money to the apparatus fund in 1880 and, as previously mentioned, gave £5000 of his £7000 Nobel Prize money to the Laboratory.

Second, unlike Rayleigh, who set up an apparatus fund, both Thomson and Rutherford regarded manpower as sufficient. Thomson wrote that he had 'any number of pupils' willing to help him. He did not appear to realize that these same pupils were frustrated and held up in their work for lack of apparatus that a little extra money could buy. R. J. Strutt (fourth Lord Rayleigh) recalled the effect of such stringency:

> The smallest expenditure had to be argued with him, and he was fertile in suggesting expedients by which it could be avoided — expedients which were more economical of money than of students' time. . . . Naturally, this financial stringency and the rapidly increasing number of workers in the Laboratory created a severe competition for such apparatus as there was. . . . Naturally the scarcity led to the development of predatory habits, and it was said that when one was assembling the apparatus for research, it was necessary to carry a drawn sword in his right hand and his apparatus in his left. Someone moved an amendment — someone else's apparatus in his left.[20]

Similarly, W. L. Bragg left a vivid account of his time as a research student:

> There was only one foot bellows between the forty of us for our glass blowing which we had to carry out for ourselves, and it was very hard to get hold of it. I managed to sneak it once from the room of a young lady researcher when she was temporarily absent, and passing her room somewhat later I saw her bowed over her desk in floods of tears. I did not give the foot pump back.[21]

And later:

> When I achieved the first x-ray reflections I worked the Rumkorff coil too hard in my excitement and burnt out the platinum contact. Lincoln, the mechanic, was very annoyed as a contact cost ten shillings, and refused to provide me with another for a month. I could never have exploited my ideas about x-ray diffraction under such conditions.[22]

Instead he relied on his father's laboratory in Leeds, where facilities were much better.

Like Thomson, Rutherford was averse to asking for money, despite repeated requests for better facilities from his students, whose accounts of working conditions are very reminiscent of Thomson's era.

3 Thomson's Leadership: Teaching in the Cavendish

I have discussed what Thomson contributed to the later 'Cavendish' style of physics. I now want to look at how he made this contribution. It immediately becomes evident that his role was largely unconscious. He made little attempt to found a research school in modern physics, or to convert students to his approach. As his reputation increased, the students just came. This is shown by the figures for the recruitment and publications from the Cavendish.[23]

Figure 6.1 shows the total recruitments of the Laboratory and the recruitments from Cambridge. As expected, there was a rapid increase from outside Cambridge after 1895, when the statutes were altered to allow graduates of other universities to take a research degree at Cambridge. However, an interesting point, which I will return to later, is that the numbers of Cambridge graduates going on to do physics did not increase very much after

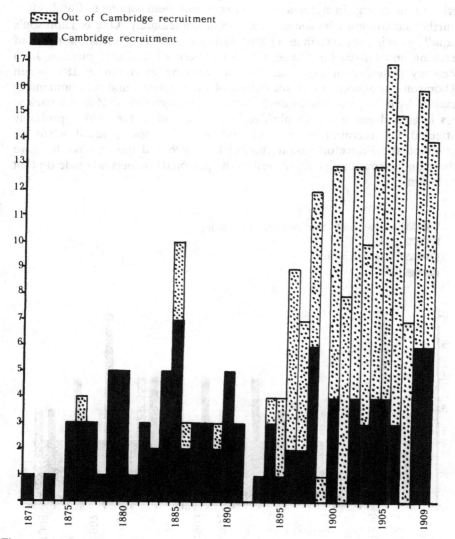

Figure 6.1 Recruitment to the Cavendish Laboratory from Cambridge and elsewhere.

1896, as it might be expected to do, in view of general public interest and excitement about physics following the discovery of X-rays and radioactivity. Perhaps enthusiasm for such modern topics did not percolate through to the undergraduate teaching at Cambridge!

Figure 6.2 breaks down the 'out of Cambridge' recruitment after 1894 into four categories: other UK universities; British Empire; USA; Europe and the rest of the world. What this graph does not show is that most visitors from Europe and the USA were not registered research students. They came on one-year sabbaticals and were often non-collegiate. There was nothing even prior to the change in statutes in 1895 to prevent them coming to Cambridge. Further factors must be sought. Two occur immediately. One is Thomson's rapidly growing reputation as a particularly imaginative physicist, a source of exciting and far-reaching ideas at the forefront of 'modern' physics. John Zeleney provides an example. He was working in Berlin in 1897 when Thomson announced his identification of the electron, and was immensely excited by it. No one else in Berlin believed in electrons, so Zeleney packed his bags and made for Cambridge.[24] The second factor, which probably affected US recruitment, was the interest Thomson aroused when he attended the Princeton sesquicentennial in 1896 and the lectures he gave there on gaseous discharge, as well as the personal contacts he made on that occasion.

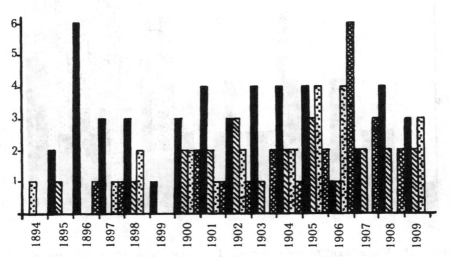

Figure 6.2   Breakdown of the 'out of Cambridge' recruitment after 1894.

The next graph, of what researchers at the Cavendish were doing (Figure 6.3), is also illuminating as to why students came to Cambridge. It shows the numbers of researchers working in different areas of physics, on the basis of the publications coming from the Laboratory: electromagnetism; standards and precision instrumentation (i.e. the type of research Rayleigh had encouraged); gaseous discharge (mainly work related to Thomson's experimental programme); everything else. The top graph shows this for Cambridge graduates; the bottom, for visitors from elsewhere. The difference is striking. Overwhelmingly, the visitors came because they wanted to work with Thomson on his exciting new discharge programme. The Cambridge graduates, however, were only slightly more likely to work on discharge than on some more traditional aspect of physics. Until he received a large influx of students from outside who were already largely committed to his ideas, Thomson's research programme attracted very few students, and he does not appear to have made any attempt to increase his recruitment.

This, again, suggests that Thomson's influence did not reach the undergraduate teaching, and this is borne out by the reminiscences which suggest that Thomson did not take much interest.[25] Most of them disguise this by saying that he was wise enough to let demonstrators and lecturers have a free hand. But I suggest that there was a more serious split between teaching and research. This is indicated by G. P. Thomson's recollection that only research workers came to the Laboratory tea, the social event of the day, around which life at the Cavendish revolved: teaching staff kept away.[26] It is shown even more strongly in the recruitment and research topic figures (Figures 6.2, 6.3). The recruitment figures suggest that excitement with the rapid developments in modern physics did not reach undergraduates at Cambridge. Perhaps the lecturers and demonstrators themselves were not excited? The research topics support this hypothesis. The graph does not give names, but if they are filled in, we find that the lecturers and demonstrators contributed almost entirely to the 'standards' and 'everything else' categories. They did not participate in the discharge research programme. They represented an old guard who had been educated in the nineteenth-century 'Cambridge' tradition and continued their teaching in this tradition. Their students, in turn, were almost as likely to choose classical research problems as modern ones.

This neglect of undergraduate teaching is one instance in which Thomson provides a contrast with his successor, Rutherford, who made strenuous efforts to encourage undergraduates at McGill into research in modern physics, and at Manchester turned the whole of the third-year teaching into a modern physics course.[27]

Thus, it is evident that Thomson made no deliberate attempt to start a research programme, as Rayleigh had done and as Rutherford did elsewhere, or to attract students to it. Research workers came flocking to the Cavendish after 1895, but they came because of Thomson's reputation; because they wanted to work on modern topics. They were already converted prior to arrival. Once there, they were fascinated by Thomson's personality — above

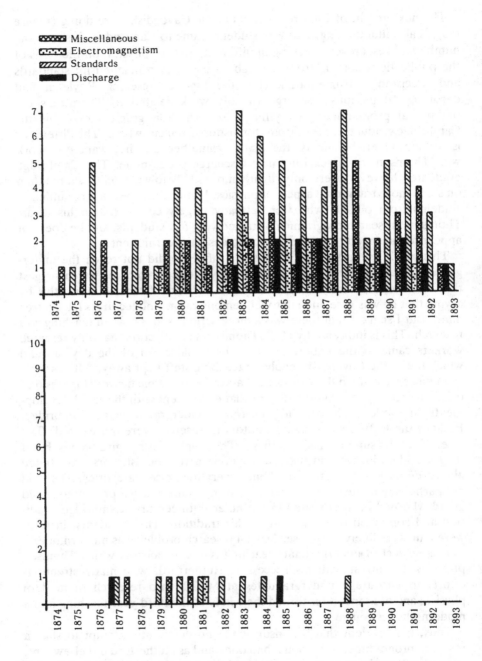

Figure 6.3 (above and opposite) Number of publications in different areas of physics coming from the Cavendish Laboratory. Top: papers by Cambridge graduates (including Thomson's own contribution). Bottom: papers by graduates from elsewhere.

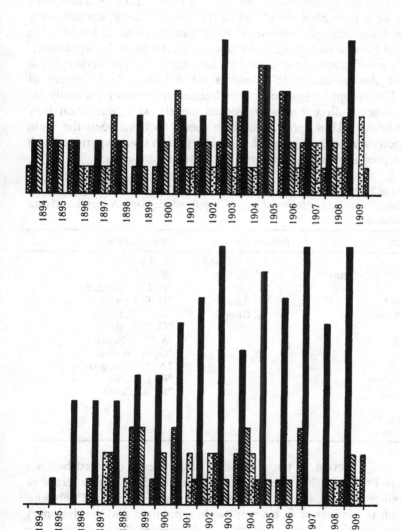

all, by his friendliness and enthusiasm.[28] He cemented their commitment to modern physics, and, above all, instilled in them his enthusiasm for physics as an exciting and stimulating subject. It was this obvious excitement with the subject that marked out many of Thomson's (and later Rutherford's) former research students.

Table 6.2 bears these conclusions out. It is a list of those of Thomson's students who were later elected Fellows of the Royal Society, divided, very roughly, into those strongly influenced by Thomson (who followed what might be called his research programme and adopted a lot of his approach), those virtually uninfluenced, and an intermediate category. The main point to notice is that the 'uninfluenced' students all did their first degrees at Cambridge! The receptive students were frequently outsiders, already inclined in Thomson's direction, who worked closely with him when they reached Cambridge. Thus, prior education seems to have been the main influence in determining their research course, although their experience with Thomson was probably largely responsible for their later success.

Table 6.2   List of FRSs educated under Thomson, divided into those strongly influenced by him, those little influenced, and an intermediate class. Asterisks indicate Cambridge graduates

Strong influence	Intermediate	Little influence
*G. A. Schott	*C. T. R. Wilson	*C. Chree
*W. C. Dampier Whetham	G. W. C. Kaye	*R. Threlfall
E. Rutherford	T. H. Laby	*H. L. Callendar
J. S. Townsend	F. W. Aston	*H. F. Newall
J. A. McClelland	*W. L. Bragg	*G. F. C. Searle
*H. S. Allen		*G. T. Walker
H. A. Wilson		*S. W. J. Smith
*R. J. Strutt		*G. W. Walker
C. G. Barkla		*A. S. Eddington
*O. W. Richardson		*G. I. Taylor
*N. R. Campbell		*E. Appleton
F. Horton		
*R. Whiddington		
*G. P. Thomson		

The style and direction of the Cavendish undoubtedly changed because Thomson was Professor, but this was not a simple effect of deliberate or dictatorial policy in the Laboratory. It was a far more vague and nebulous affair in which the influence of his worldwide reputation and revolutionary theories, coupled with increased school and undergraduate teaching in physics, played a large part, as did the unconscious effect of his personality.

Notes and References

1. For a description of the original Laboratory buildings, see *Nature, Lond.* (1874), **10**, No. 243, June 5, 139–42. For descriptions of the later buildings, see Crowther (1974).
2. Thomson and Tait (1912), pp. 442–3.
3. Schuster and Shipley (1917), pp. 12–13.
4. See, for example, Rayleigh (1899–1920).
5. Discussed in detail in Harman (1985).
6. *A History of the Cavendish Laboratory 1871–1910*, London, 1910.
7. Hendry (1984).
8. E. Rutherford to A. S. Eve, quoted in Eve (1939).
9. Thomson's approach to experiment is discussed in detail in my forthcoming paper, 'J. J. Thomson: an experimental genius?' A case study of how this approach worked out in practice in Thomson's investigation of positive rays is contained in Falconer (1988).
10. Rayleigh (1942), p. 18.
11. Falconer (1988).
12. Schuster (1911), p. 52.
13. Schuster (1884), p. 318.
14. It may be significant that Thomson had attained a secure and highly esteemed position before embarking on such unpopular research.
15. Falconer, forthcoming, *op. cit.* [Note 9].
16. Rayleigh (1942), p. 24.
17. These problems are emphasized by Buchwald (1985), especially pp. ix–xiii, 3–4.
18. *A History . . .*, *op. cit.* [Note 6], pp. 279–80; Thomson (1936), p. 124.
19. J. J. Thomson to O. Lodge, 4 December 1905, Birmingham University Library, Lodge papers, OJL 1/404/20.
20. Rayleigh (1942), pp. 48–9.
21. Phillips (1979), p. 84.
22. *Ibid.*, p. 90.
23. These figures are drawn from the lists of researchers and publications in *A History . . .*, *op. cit.* [Note 6], pp. 281–334.
24. Jaffe (1952).
25. Rayleigh (1942); Thomson (1964); obituaries in *Nature, Lond.* (1940), **146**, 351–7.
26. Rayleigh (1942), pp. 53–4.
27. Robinson (1954), p. 10; Wilson (1983), pp. 180–1.
28. Obituaries, *op. cit.* [Note 25].

7

Astronomical Observatories as Practical Space: The Case of Pulkowa

Mari E. W. Williams

Astronomical observatories are among the oldest of scientific institutions. By the end of the seventeenth century in Britain and France, national resources had been invested in the study of astronomy, resulting in two of the most famous observatories in the history of the subject: those at Greenwich and Paris.[1] For over a century these two institutions dominated developments in observational astronomy, but by the early nineteenth century they had been joined by similar, if sometimes smaller, observatories all over Europe. Astronomy by this time occupied a highly favoured place in the minds of many European natural scientists; the power of Newtonian and subsequently Laplacian and Lagrangian celestial mechanics to predict the positions of astronomical bodies and to explain much about the nature of solar system objects gave the discipline so firm a basis as a continuously testable, predictive study that it was regarded as a model science.[2] At the same time, the early decades of the nineteenth century witnessed important changes in the study of astronomy. The subject was being led in new directions, particularly by astronomers from the German states; individuals such as Friedrich Bessel, Wilhelm Olbers, C. F. Gauss, Johann Encke and Wilhelm Struve were rewriting the ground rules of astronomical practice, and they came to be deferred to as the leaders of the subject throughout the Continent. Their work constituted the acknowledged forefront of astronomical endeavour, and their methods of working were adopted throughout Europe and, subsequently, the United States and the European colonies.[3]

However, the details of how these people worked, as opposed to the results they produced, have not been explored to any great extent by historians. This chapter will begin such an exploration; in particular, it will address the questions of the spaces in which astronomical activity took place. For instance: Which work was carried out within the observing rooms and which elsewhere? How were observatories equipped? How were sites for observatories chosen? How were they funded? Finally, questions related to the buildings themselves: What were they meant to represent? Who was allowed to use them or to visit them? How was the space within them allocated and used?

For this preliminary investigation the one observatory which will be considered in detail is the Russian Imperial observatory at Pulkowa, near St. Petersburg. During the second half of the nineteenth century this observatory was regarded widely as the most significant and complete astronomical observatory in the world; it was described variously as the 'astronomical capital of the world',[4] an 'astronomical palace' and an 'astronomer's inspiration'.[5] For several decades this observatory was the home of some of the most exciting developments in astronomy; and at the same time it was the source of the highest standard of routine astronomical observing, the production of fine astronomical tables and one of the best libraries of astronomical works in the world.[6] By examining the observatory from the points of view of its physical structure and of the organization of work within it, therefore, it will be possible to understand an important aspect of how such admirable results were produced. Moreover, such an investigation will shed light on how, by the mid-nineteenth century, astronomy was, at its most organized and expensive, a joint, institutionalized activity.

The Pulkowa observatory was opened with a grand ceremony during the summer of 1839, but plans for the new establishment had been in hand for over a decade, and the idea of a new Imperial observatory had surfaced from time to time for far longer. In certain respects, the fact that the most revered of astronomical observatories in the mid-century was to be found in Imperial Russia is curious; as far as astronomical activity was concerned, the leading practitioners during the 1820s and 1830s were concentrated in the German states around Göttingen, Berlin, Seeberg, Altona, Bremen and Königsberg, and Russia had little representation in the previous history of astronomy. But the opening chapters of the detailed history of the foundation of Pulkowa set its early development firmly within the context of the history of science in Russia, portrayed as a specifically tsarist endeavour.[7] Peter the Great was credited with having introduced science to Russia, founding the St. Petersburg Academy of Sciences in 1724 and establishing an astronomical observatory in the city a year later. The French astronomer Joseph Nicholas Delisle was invited to be the observatory's first director, and under him and equally eminent successors nationally backed programmes of geodetical survey and of astronomical observation of such important occurrences as the two transits of Venus in the 1760s were carried out.[8]

By the turn of the eighteenth century, however, there was little serious astronomical activity at the St. Petersburg observatory itself, which was badly sited in the centre of a busy town. The most important observatory in the Russian Empire during the early nineteenth century was that attached to the University of Dorpat in Estonia. But the idea that St. Petersburg should have its own, internationally significant observatory persisted, and when, in 1824, the directorship of the city's observatory became vacant, the idea resurfaced in a sufficiently concrete form:

> The founding of a new academic observatory became urgent, as it was appreciated that the Academy must preserve its dignity with respect to practical astronomy, its dignity as the premier scientific institution of the Empire, a dignity invested in it by its illustrious founder, Peter the Great.[9]

From the outset, then, the proposed observatory was to be special; it was to reflect the status of the St. Petersburg Academy of Sciences and, beyond that, to represent science in Imperial Russia. With such an ambitious aim the backing of the Government was inevitably sought and, once the Tzar, Nicholas I, had endorsed the plan with promises of Treasury funding, the Academy was able to start the process of creation. In charge of the plan was Georg Friedrich Parrot, Professor of Physics at the University of Dorpat, who had previously been responsible for the housing of the famous Fraunhofer refractor at the Dorpat observatory in 1834.[10]

The acquisition of the Fraunhofer instrument and the uses to which it was put were masterminded by the Director of the observatory, German-born F. G. W. Struve. Struve had joined the staff of the observatory in 1813 after taking his doctorate from the University of Dorpat, and six years later he became Director.[11] As such, he occupied the most senior astronomical post in the Empire, and it is hardly surprising that he was approached to take on the job of Director of the new Imperial observatory. In 1830 he was appointed and put in charge of equipping the observatory and of overseeing the choice of its design.

Once the early decision regarding the appointment of Struve was made, it was some time before the money for equipping and building the observatory materialized. However, by the spring of 1834, when a Commission under the direction of Admiral Alexis Greig was established, much had been accomplished. Two architects had been invited to submit plans for the buildings, and substantial sums of money had been made available for the buying of instruments. On receipt of the respective sets of plans, from Monsieur Thon and from Alexandre Bruloff, the decision turned out to be easy, although the reason for the choice is important. Thon had submitted plans which were, according to the Commission, suitable for a pleasant country residence, whereas Bruloff had met the specific aims of the Commission for the new observatory.

> M. Thon's façade in the Gothic style had the advantage from the point of view of architectural beauty over that of M. Bruloff, which had no pronounced style, but which on the other hand, clearly indicated through its special character, the scientific role of the building.[12]

At first glance Bruloff's building looked like a great observatory; moreover, his design was based upon an appropriate arrangement of spaces within the observatory, with study rooms placed near observing rooms and direct links between working and living areas. Finally, Bruloff's plans featured lower towers for the main instruments than did those of his rival, which would mean greater rigidity for the support of those instruments. Bruloff's plans were therefore approved by the Commission and presented to the Emperor as the better of the two proposals; with the necessary Imperial commendation the plan could be formally adopted.

But before this part of the plan could be acted upon, a suitable site had to be found. Obviously, the centre of St. Petersburg would not do; a clear and wide southern horizon was vital, so that the offer made by Academy member Count Kouchelev-Besborodko to use part of his estate had to be rejected because his land lay to the north-east of the city. Eventually, the Tzar intervened again, instructing his Minister for Public Instruction to find a site, and when the village of Pulkowa, some 12 miles to the south of St. Petersburg, was suggested, it was Nicholas who authorized the building of the observatory there and who issued an Imperial decree forbidding any other new buildings in the vicinity.

While final adjustments were made to the architectural designs and construction work was started, Wilhelm Struve, armed with a very basic plan of the observatory, travelled across Europe shopping for the best instruments money could buy. His instructions were that the Emperor 'wished to see the new observatory supplied with everything of the most perfect' as far as instrumentation was concerned.[13] Struve's favoured instrument makers were all German: at Munich, where Ertel had succeeded Fraunhofer; just outside the same town, where Merz and Mahler had inherited Reichenbach's workshop; and the Repsold brothers in Hamburg.[14] His decision taking was aided by lengthy discussions with his peers in the German states, including Bessel, Encke, Humboldt, von Lindenau and Olbers, with each of whom he spent time on his travels. He was, moreover, clearly predisposed towards the Munich shops, as the main, and highly successful, instruments he had acquired at Dorpat had been supplied by Fraunhofer and Reichenbach. Negotiations between Struve and the makers were prolonged, and, after his return to St. Petersburg, it proved simpler to send an astronomer from the Academy back to Munich than to continue via an involved correspondence. The completed instruments were eventually transported to Pulkowa in the early summer of 1839 and, in keeping with Bessel's precept that all instruments are built twice (once in the workshop and again when they are assembled and tested *in situ*[15]), it was another few weeks before the observatory was considered to be ready and the opening ceremony was arranged.

But what was it that was opened? What did the observatory look like? Who worked there? And what did they do? Figure 7.1 shows the general scheme of the observatory and its grounds viewed from above. The village of Pulkowa,

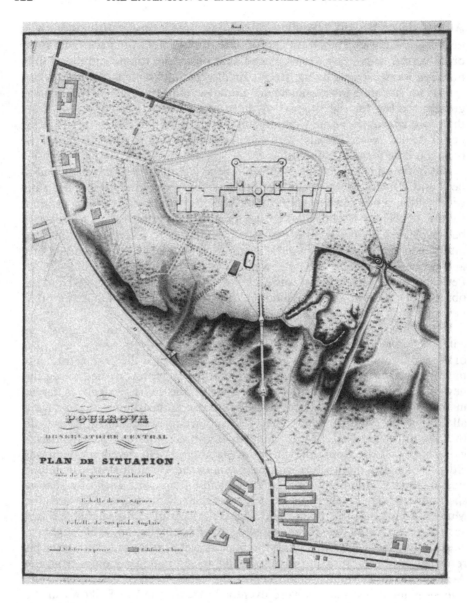

Figure 7.1 Pulkowa Observatory. Plan de situation. From Struve (1845), Vol. 2, Planche I.

immediately to the north of the observatory, is shown at A, at a point of division on the main road from St. Petersburg (B), and to the south of the observatory is the wide, empty horizon so necessary to successful observing. Not apparent from this view is the steady rise from Pulkowa village to the observatory, sited at the top of a hill, which meant that the lightly wooded areas within the observatory grounds did not pose any problem. The grounds of the observatory extended northwards almost to the village and, to the south, within an arc of approximately 200 yards radius, with pedestrian access very near the village, and access for vehicles to the east of the main buildings, at E. Thus, the overall situation of the observatory had the advantage of being near enough to a main route south from St. Petersburg to maintain necessary supplies and to make regular visits to the town feasible; but, at the same time, it was far enough away from that metropolis to deter the 'often too frequent' visits of 'the curious'.[16]

The isolation of the observatory was an important feature. Members of the Commission and, especially, Struve were of the opinion that a sense of community should be encouraged among the observatory staff. They should create between them 'an intimate and pleasant circle, fostered by the common interest inspired in them by the sublime science they pursued'.[17] Whereas a number of new scientific institutions created across Europe during the first half of the nineteenth century were public places, where scientific discoveries and results were presented to a wide audience, astronomical observatories, and particularly Pulkowa, were private spaces from which the results were generated.[18] Astronomers not only worked there, as will be discussed more fully below, they also lived there for months at a time and all their needs were provided for within the grounds occupied by the observatory. The gardens shown in Figure 7.1 (p, q, r, s, t, u, v and w) were for the use of staff with the proviso that the hierarchy of the institution was respected: staff of different status had access to different parts of the garden. Moreover, the site of the observatory was self-sufficient in terms of water and was, overall, 'one of the most advantageous that could possibly be found.'[19] Even the possible problem of severe weather conditions was considered as dealt with: a row of young trees was planted to the south-west of the main buildings to serve as a windbreak, and also to help prevent too much snow from piling up against the observatory walls.

Within these supposedly ideal surrounds, the main work of the observatory was carried out in the central buildings, the ground plan of which is shown in Figure 7.2. The central part of the observatory and the focus of all activity was made up of the structures surrounding A, the central observing space; the outer small observatories at c, c', c'' and d completed the observing parts of the establishment. With the working spaces in their rightful central positions, the other parts of the institution were arranged symmetrically to the east and west. At the ends of the two corridors D and D' lay the residential areas, with the Director's house at E on three floors, including a basement. The single-storey building G housed one assistant astronomer, with three others, the inspector and the technician living in buildings H and F, mirror images of G and E except for having an extra floor allowed by the lower terrain to the east of the central observatory. From such an arrangement it was evident that at least the senior staff of the observatory were expected to regard Pulkowa

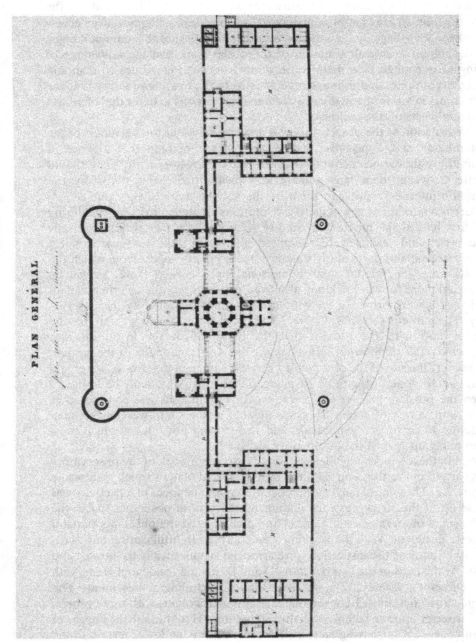

PLAN GÉNÉRAL

Figure 7.2 Pulkowa Observatory. Plan général. From Struve (1845), Vol. 2, Planche IV.

Observatory as their home, with all their needs catered for through the provision of stables, coach houses, ice houses, wash houses, bathrooms and a blacksmith, located at I, I', K, K' and L. Furthermore, facilities allowed some families to be accommodated.

Because of the needs of astronomy, all the buildings attached to the central observatory had to be constructed to very particular standards. They were solidly built in brick, on a foundation of limestone blocks covered with iron sheets. All the external walls were made 3 feet thick in an attempt to keep the ambient internal temperature, maintained by stoves, as steady as possible, even during the extremely cold winters; and, with the prevention of internal temperature gradients also in mind, the stone stairwells were heated.

In brief, all the living areas were extensive and they united the arrangements necessary to deal with a site as exposed as that of the central observatory, with those needed for everyday life.[20]

The observatory itself was 'a building of purely scientific significance', which had to include: rooms for making observations, with different ones for the large principal instruments from those needed for the smaller portable ones; a reception room; offices for the Director and for four assistant astronomers to be placed conveniently for their observing commitments; several smaller offices for visiting savants, who were expected to stay at the observatory on a temporary basis to pursue individual programmes in practical and theoretical astronomy; rooms for keeping and maintaining auxiliary equipment and portable instruments; a special room for examining the effect of temperature on the observatory's clocks; and, finally, a place for a library.[21] On the observing side, the statutes of the institution stated that there must be two distinct spaces: first, for the large, permanent instruments to be used for the principal work of the observatory 'to perfect astronomy as a science'; and second, room to erect portable instruments which could be used both for individual observing programmes and for astronomers to practise their skills with the aim of perfecting 'the art of observation'.[22] It was because of the incompatibility of these different activities that the four small outer observatories (c, c', c″ and d) were built, leaving the main observatory to house the main instruments.

These latter instruments were themselves divided into two categories: those which were to remain in fixed planes, either on the meridian (north–south) or the prime vertical (east–west), and to be used for the fundamental task of plotting stellar positions; and instruments which moved freely in azimuth and, hence, were used for a range of astronomical studies, and instruments used for the detection of parallaxes.[23]

In designing the central observatory, then, Bruloff had to create a building which allowed these different instruments to function freely. The results of his deliberations can be seen in Figures 7.3–7.6. Figure 7.3 shows the northern and southern facades of the building, all aspects of which, according to

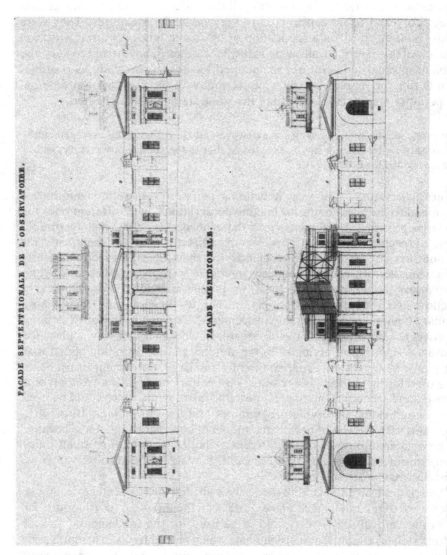

Figure 7.3 Pulkowa Observatory. Façade septentrionale; façade méridionale. From Struve (1845), Vol. 2, Planche V.

Struve, 'illustrate a particular role for the building and emphasise the scientific character which distinguished it'.[24] Of particular importance, of course, was easy access to the sky, and the facades show the high central dome covering the main Merz and Mahler refracting telescope, and the slits in the roofs of the east and west observing rooms at E and D, respectively. The Merz and Mahler instrument was free to move in azimuth and altitude under a rotating tower, whereas the slits over the various instruments in rooms E and D show that these instruments were fixed in the plane of the meridian.

Inside the building it is clear from Figure 7.4, showing the ground floor, that the entrance to the observatory was designed to impress. The observatory was approached via the stone steps on the north side (a) leading to the peristyle (b) and then the entrance hall (c). Visitors to the observatory would then find themselves in an imposing space surrounded by pillars apparently supporting a domed roof (G); the purpose of this space and the pillars was actually to support the floor of the room containing the Merz and Mahler refractor, but they could nevertheless be used to frame an impressive reception area, emphasizing the significance and size of the activity undertaken at Pulkowa. Of the eight spaces between the massive pillars, four — ε, ε', ε'' and ε''' — were filled with mahogany rails where numerous portable instruments were kept, as well as a barometer. These, together with the clock placed in the niche η, would have conveyed to all the functional nature of the building they had entered, but at the same time the grandeur of the establishment was reinforced by the many portraits hung around the walls: images of Copernicus, Tycho, Kepler, Newton, Flamsteed, Halley, Bradley, Roemer, William Herschel and Schubert represented the previous achievements of astronomy, while the honours of the contemporary discipline were shared between, among others, Airy, Bessel, Ertel, Gauss, John Herschel, the Repsold brothers, Schumacher and James South. Finally, the whole was overlooked by a marble bust of Emperor Nicholas I, strategically placed in the embrasure ξ opposite the entrance to remind everyone by whose grace the institution had been created.

Beyond the central room, to the south, lay the library (l), which, by 1845, already boasted 5000 quarto volumes, and an observing room (F) housing a Repsold transit instrument which was kept in the plane of the prime vertical and thereby complemented the three meridian instruments in rooms E and D (at d, e and g). E, the eastern room, housed a Repsold meridian circle, while the main meridian instrument in D, the western observing room, was a vertical circle made by Ertel. On either side of the main observing spaces were the astronomers' offices, where they would work on the reduction of the data gathered during spells of observing. Struve, as Director, occupied the large room, r, slightly separate from the rest of the staff, as befitted his status, which was further underlined by the portraits adorning his walls, including paintings of Serge Ouvaroff, Minister for Public Instruction and President of the St. Petersburg Academy of Sciences; Admiral Greig; Monsieur le Prince Michel Dondoukoff-Korsakoff, Vice-President of the Academy; and Paul-

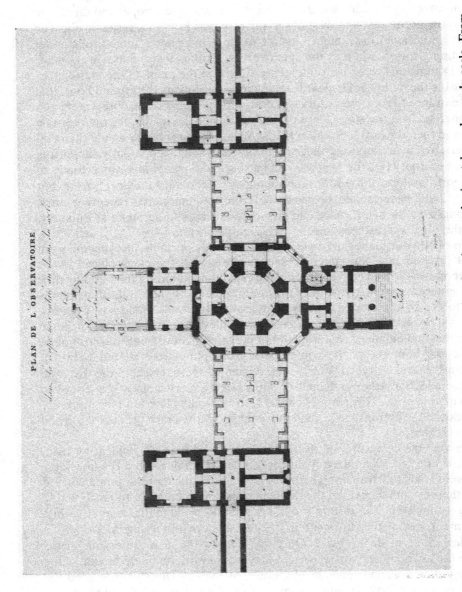

Figure 7.4 Pulkowa Observatory. Plan d'observatoire: dans la coupe horizontale au dessus du socle. From Struve (1845), Vol. 2, Planche VII.

Henri Fuss, its Secretary. These portraits also served to remind the Director, should that be necessary, of the closeness between the Academy and the Imperial family and the roles of both in the founding of the observatory.

To maintain the symmetry of the structure there was a room of equal size to the Director's office, at r', but, because of the need to sustain the hierarchy of the institution, this could not be used as a working space of similar nature to Struve's room. It was, instead, used to store equipment not used regularly, while the four resident astronomers worked in the offices on the north side of the building at u, v, u' and v', depending on which of the two rooms D and E the astronomers were placed in for their observing work. Apart from these rooms, the final space of interest was that in which the observatory's many clocks were checked, particularly for their reactions to changes in temperature; this was located at y, near the front entrance, completing the allocation of space at ground level.

Three stairways, at z, α and α', led to the first floor, illustrated in Figure 7.5. The most significant feature of this level was the central room containing the large refractor. This space, the focal point of all the observatory's activity, was characterized by two obvious features: the solid support for the telescope and the moving tower. The base of the instrument was set in masonry, lying immediately over the domed ceiling of the reception room described above; rising vertically above the eight pillars which supported the dome was a cylindrical wall, on top of which the rotating tower was placed, the motion of which along a circular track was controlled by a system of pulleys and wheels. The instrument itself was supported at the level of the base of the tower (the level of the balustrade shown in Figure 7.3) and was reached by two staircases from the first floor (d and e in Figure 7.5). When in use, the telescope was pointed to the sky either between the two halves of the tower, for low-altitude observations, or through the roof, for higher altitudes.

Elsewhere on the first floor of the building were two further observing towers, at B and C; by 1845 the space B was still empty, but was destined to house a telescope, to be used for parallax measurements, while C was the site of the observatory's Merz and Mahler heliometer. This instrument was used for measuring small celestial angles such as, for example, the separation between the components of a double star, or the diameter of celestial objects, originally the Sun, whence the instrument's name.[25] Rooms B and C were reached via separate staircases — as there was no connection between them and the central tower — and the construction of the two towers was a scaled-down version of the main one. These observing spaces were clearly of less significance to the overall functioning of the observatory than the central tower and the ground floor observing rooms. Moreover, in accordance with the hierarchical arrangement of the building, the offices on this level, to the north of B and C, were allocated to visiting astronomers.

Possibly the most important aspect of the first floor, away from the Merz and Mahler telescope, was the access it accorded for maintenance of the roofs of the ground floor observatories. The three roofs in question were linked by

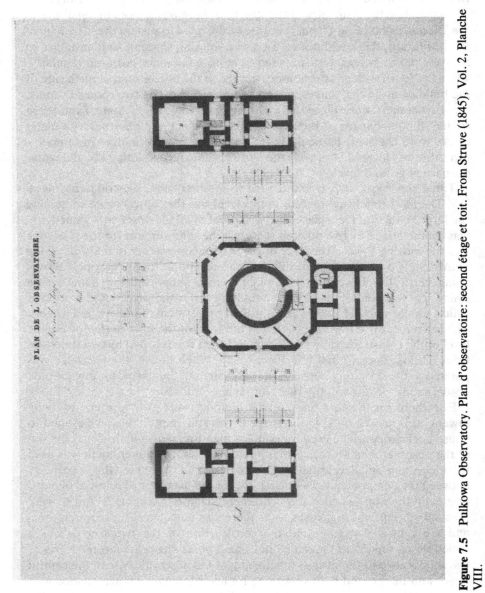

Figure 7.5 Pulkowa Observatory. Plan d'observatoire: second étage et toit. From Struve (1845), Vol. 2, Planche VIII.

outside galleries around the outside of the central tower, and were reached from the inside gallery. Over the east and west observatories, where so much of the basic positional astronomy was carried out, were planked sloping roofs over painted linen drapes; each of the trap-doors opened during observing sessions consisted of two layers of planks mounted on windlasses. They could be opened with ropes and closed with the help of springs and their own weight. Failure to close the trap-doors properly could result in serious damage from exposure of the instruments to bad weather, so that the maintenance of the observatory roof constituted an important part of the technical staff's routine work.

If the roofs of the observatory had great significance for the proper functioning of the establishment, that of the basement was no less obvious. Figure 7.6 shows this level in some detail. It was divided into three separate sections isolated by walls (x) and accessible via staircases a, b and c, respectively. In each section the most striking features were the massive stone foundations for the main fixed instruments (marked d and e). Rigidity and stability — for the building as a whole, as well as for individual instruments — were crucial for the precise observing which formed the core of the astronomer's routine, and these started with good foundations. Beneath the east, west and south observing rooms, then, in addition to the foundations for the instruments and for the external walls, there were extra, interior wall foundations. Moreover, under the central observing tower the heavy pillars supporting the Merz and Mahler refractor had deep foundations, which also contributed to the overall stability of the building as a whole.

It was not only for structural reasons that the basement was vital, however. Once the building and instruments were given as firm a basis as was physically possible, astronomers had next to concern themselves with maintaining consistent environmental conditions, and the one over which they had greatest control was ambient temperature. The basement played its part in this by being the location of the central heating boilers. At the Pulkowa observatory there were three (at l, l' and l''). From l air ducts fed warm air into the test room (y on Figure 7.4) immediately above, into the reception-room (at ι and θ), into the vestibule (c) and to a number of outlets on the first floor. In similar ways, via air ducts in the walls, l' and l'' heated the two wings of the observatory. Protection of the building from predictable changes in temperature was attempted with, for example, a shield which could be opened over the roof of the south observatory during periods of strong sunshine. Beyond that, through the use of heavy double doors, wherever possible, unwanted draughts were excluded, and all staff working at the observatory had to regard the building as a whole as an astronomical instrument.

Through its architecture, then, the observatory at Pulkowa represented a number of different things. It was, on the one hand, a statement of Imperial commitment to a certain type of scientific endeavour: orderly, hierarchical and expensive. The number of towers and observing spaces set it apart from

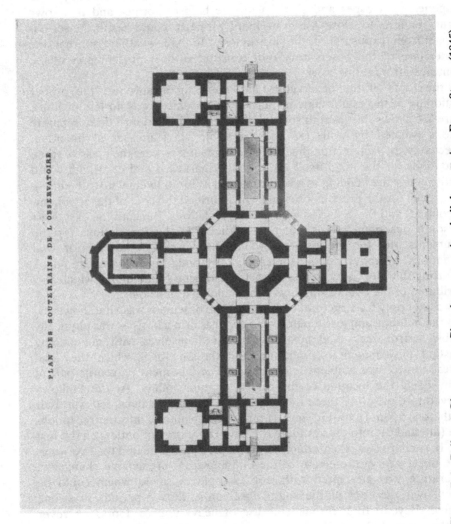

Figure 7.6 Pulkowa Observatory. Plan des souterrains de l'observatoire. From Struve (1845), Vol. 2, Planche VI.

all other observatories then in existence, and its lavish endowment of instruments and decorations bore witness to its importance as a highly significant state institution. On the other hand, the observatory was a comment about the forefront of astronomical activity in the late 1830s. As already mentioned, during the early decades of the nineteenth century, a number of astronomers were addressing very particular problems in their programme of astronomical reform; their efforts were devoted to problems of precision, in terms both of making accurate observations and of manipulating the data thus generated in precise ways. For the first of these, the construction of the Pulkowa observatory embodied the accumulated ideas of contemporary astronomers and certain of their predecessors in the most concrete of ways. Following, in particular, publications by Bessel in 1818 and 1830,[26] making accurate observations involved in the first place paying close attention to the physical phenomena which were known to affect the accuracy of the observations: hence the strong stone foundations maintaining the rigidity of the buildings and of the telescopes, and the enormous efforts made to regulate prevailing temperatures. Second, for effects which could not be controlled outright, they had to be monitored closely and the impact of their variation had to be measurable; thus, the observatory was supplied with a test room equipped with a battery of instruments of thermometry and barometry.

When it came to the manipulation of data, the observatory building again offers clues. Astronomers were expected to devote much of their time to the process of data reduction, during which raw observations would be transformed into information useful to the astronomical community more widely; this work was meant to be done quickly and efficiently. The close proximity of offices to observing spaces encouraged individuals to pay attention to both aspects of the work, and emphasized their interconnectedness. Moreover, facilities such as all possible sets of astronomical tables were easily available from the library, and astronomical calculators were employed to help with the mundane side of data reduction. Overall, then, this made access to the outside world unnecessary, and here again the isolation and self-sufficiency of the establishment are relevant. The observatory complex in this context again spoke through its physical structure. By building domestic facilities in addition to all possible astronomical ones, the communal and collective aspects of astronomy were enforced and, at the same time, the privateness of the community at Pulkowa was underlined.

But what was it that this private community was engaged upon, and how did the resident astronomers carry it out? The statutes of the observatory, drawn up in 1838, made at least the early programme clear—the observatory was: (a) to supply series of observations, as perfect as possible, aimed at improving astronomy as a science; (b) to provide observations necessary equally for geographical endeavours within the Empire and for scientific journeys in general; and, finally, (c) to co-operate in every way possible in perfecting practical astronomy in its applications to geography and navigation, and to offer to those persons who wished to benefit from it the chance to involve themselves in geographical surveys.[27] Within this programme, the strong emphasis on geography and navigation is unsurprising; they, after all, were regarded as normal activities for a national observatory and had been the reasons behind the creation of the earliest state-backed astronomical observatories. Nor did Struve take his duties lightly. He supervised a major project to measure an arc of the meridian from Ismail on the Danube to Fugelmaes on the Arctic coast. In addition, he was involved in international survey work: measuring the difference in latitude between Pulkowa and Greenwich, taking Altona as a convenient half-way point, and collaborating with the British Astronomer Royal, George Airy, over the measurement of an arc of latitude between Valentia in Ireland and Ursk in the Urals. Central to any surveying project was knowledge of the latitude and longitude of the observatories involved, and as a result, at Pulkowa, observations had to be analysed and reanalysed regularly. Such work formed part of the observer's routine gathering of stellar positions with the fixed instruments in the east, west and south observing rooms, and the routine reduction of these data.

Apart from such basic work, astronomers at Pulkowa were engaged on one of two areas of enquiry: solar system studies or stellar astronomy. Solar system observations were subdivided into three groups: tracking the changing position of the Sun; studying the positions of the Moon and planets; and observing comets. Comets, in particular, were of consuming interest to astronomers during the 1830s and 1840s.[28] In studying the solar system, astronomers at Pulkowa were not setting themselves apart from their peers at any other observatory; but by placing stellar astronomy firmly on the agenda, Struve was quite consciously creating a new role for a major observatory. As he put it:

> In the majority of astronomical establishments, observations of the celestial objects of the solar system constitute the main object of activity. . . . Stellar astronomy can only advance further through the use of enormous optical powers, and. . . . The powerful instruments at our Observatory· guarantee a distinguished success in this branch of astronomy, it having been decided that the main purpose therein must be to work for the advancement of stellar astronomy. . . .[29]

Thus, Pulkowa was, for Struve, a statement about the future of nationally funded, large-scale astronomy; and he was keen to pursue his plans. During his directorship both he and a number of his assistants turned their attention to the realm of stellar astronomy, thereby giving that branch of the subject an institutional grounding it had previously lacked.[30]

It is clear, then, that the observatory at Pulkowa represented the cutting edge of state-backed astronomy during its early existence. In no way, therefore, can it be regarded as an average astronomical observatory. Nevertheless, as a highly influential centre of astronomy, the design and research programmes of which proved of such interest to contemporaries,[31] it is a revealing place at which to begin an investigation of astronomical architecture. Moreover, the wealth of detail about the buildings and the arrangement of work within them left by the first Director provide fascinating insight into his and his contemporaries' views of what an astronomical observatory should be. Virtually unhindered by financial constraint, the Commission involved in the establishment of Pulkowa was able to create an astronomer's dream.

But it should not be forgotten that, while money was no object, the observatory, its buildings and equipment were governed by the fact that the Russian Emperor had sanctioned it, and would expect it, therefore, to represent Imperial interests. Overall, then, the Pulkowa observatory must be seen as a combination of the concerns of astronomers, which dictated its location, its structure and part of its activity, and the concerns of Empire, which influenced its grandeur, the convenience with which it could be built and the remainder of its programme of work.

Notes and References

1. Forbes (1975); Wolf (1902).
2. See, for example, the discussion of astronomy in Herschel (1830), pp. 265–81.
3. Aspects of these changes are discussed in Williams (1981), Chap. 4.
4. According to American astronomer Simon Newcomb, this was the view of Benjamin Gould. See Newcomb (1903), p. 309.
5. From the unpublished autobiography of Cleveland Abbe, quoted in Reingold (1964).
6. See Krisciunas (1988), Chap. 5, and the references to the history of Pulkowa cited therein.
7. Struve (1845), Vol. 1, p. 6.
8. *Ibid.*, pp. 6–13. See also Woolf (1959).
9. Struve (1845), Vol. 1, p. 22.
10. Struve (1825).
11. An account of Struve's early life may be found in Batten (1977).
12. Struve (1845), Vol. 1, p. 28.
13. *Ibid.*, p. 29.
14. On nineteenth-century German instrument makers, see Brachner (1985).
15. Bessel (1848), p. 432.
16. Struve (1845), Vol. 1, p. 73.
17. *Ibid.*, p. 80.
18. For a fine account of the architecture of some of these institutions in Britain, see Forgan (1986).
19. Struve (1845), Vol. 1, p. 80.
20. *Ibid.*, p. 88
21. *Ibid.*, p. 89.
22. The statutes are listed in *ibid.*, pp. 57–67.
23. Wilhelm Struve was among the first astronomers to be successful in measuring the elusive parallax (and hence distance) of nearby stars, and intended to pursue his interest in this further while at Pulkowa. See Williams (1981), Chap. 6.
24. Struve (1845), Vol. 1, p. 94.
25. On the history of the heliometer, see King (1955).
26. Bessel (1818, 1830).

27. Struve (1845), Vol. 1, p. 56.
28. A survey of the published material from the Royal Astronomical Society in London during the 1830s and 1840s shows a remarkable number of papers dealing with the subject of comets. The Society had set itself up as a clearing-house for this study, illustrating the continued interest of astronomers in these objects.
29. Struve (1845), Vol. 1, pp. 250–1.
30. See Struve (1847).
31. Links between Pulkowa and, especially, Harvard are discussed in Hetherington (1976).

8

The Geopolitics and Architectural Design of a Metrological Laboratory: The Physikalisch-Technische Reichsanstalt in Imperial Germany

David Cahan

Metrological laboratories — that is, laboratories devoted principally to establishing physical standards and making physical measurements — are a modern phenomenon. Although their emergence was in some cases due largely to the efforts of specific individuals, towards the end of the nineteenth century at least two sets of historical conditions had evolved to make such laboratories increasingly welcome, if not necessary. First, starting in the 1830s, physicists, especially German physicists, began seeking increasingly precise observational and experimental data that both revealed new physical phenomena and tested the explanatory power of physical theory. As a result, the new field of precision or measuring physics began to emerge. Second, and closely tied to this first point, after 1870 the field of precision physics experienced increased demands from both the physics community and industry (in particular, the old mechanical and the young electrical and optical industries). Both academic physicists and industrial interests wanted precision physics to conduct precision-measuring research aimed at establishing universally trustworthy physical standards, at creating or improving scientific instruments and apparatus, and at testing instruments, apparatus and materials.

The high costs of constructing and equipping such metrological laboratories has meant that even in the late twentieth century there are only a small number of them throughout the world and that these have had to be founded and operated by national states. To one degree or another, these laboratories — such as the National Physical Laboratory in England and the National Bureau of Standards in the United States — have been modelled on the first national laboratory devoted to research in physical metrology and the testing of physical instrumentation and apparatus: the Physikalisch-Technische Reichsanstalt (PTR) or, to use its English-language title, the Imperial Institute of Physics and Technology.

The key concern in planning and designing metrological laboratories is the environmental control of both external and internal sources of disturbance to the measuring process. Prior even to the need to work with the most precise measuring instruments and the highest-quality apparatus is the assurance, for the measuring physicist, that the physical laboratory in which he conducts his measurements provides the utmost in such control. Significant disturbances from beyond or within the laboratory are intolerable; his workplace must be a setting that eliminates or at least minimizes possible disturbances.

Environmental control of external sources of disturbance is a matter of what I shall call *geopolitics*, by which I mean the geographical and political setting of the metrological laboratory. The planners of the laboratory must find a means of securing it from potentially disturbing features of its physical (including geological) environment and location, from the expected be-haviour of the surrounding population, and from the existence of other nearby institutions or facilities, such as industrial firms and transportation facilities. Further, the planners must provide a buffer zone between the laboratory and its surrounding environment, so as to allow for potential growth of the laboratory itself as well as future changes in the environment which might subsequently disturb the laboratory's work. Environmental control from within, for its part, is a matter of *architectural design*. Here planners must design the laboratory in such a way that internal mechanical, thermal and electromagnetic effects do not disturb the laboratory's delicate metrological work, and so that there is flexibility in the building's internal structure which will accommodate changes in the type of research and testing work conducted by the laboratory.

Below I show how the planners of the new Reichsanstalt secured environ-mental control for their laboratories between 1887 and 1901 by scrupulously addressing these matters of geopolitics and architectural design. Elsewhere I have discussed the related issue of the Reichsanstalt's founding in 1887 and the central role played in that founding by the industrial scientist Werner Siemens.[1] In this chapter I shall restrict my discussion to the location, layout and design of the Reichsanstalt's physical plant, showing how the various decisions taken by the Reichsanstalt's planners sought to obtain the environ-mental control necessary for conducting metrological and testing work of the highest order then possible.

1 The Geopolitics of a Physics Institute: Berlin–Charlottenburg, Werner Siemens and Electric Streetcars

From the initial efforts in the 1870s to found a precision mechanical institute to the latter's final transformation into the far more ambitious Physikalisch-Technische Reichsanstalt, with its bifurcation into a Physical Section (*Wissenschaftliche Abteilung*) and a Technical Section (*Technische Abteilung*), there had never been any doubt that the new institution would be located in Berlin.[2] That was true both because the Reichsanstalt's principal founder, Werner Siemens, and his electrical firm of Siemens & Halske, like many other electrical firms, was located in Berlin and because the political leadership of the Reich sought to make Berlin into the centre of Germany's political, economic, cultural and academic life. After the Reich's founding in 1871, its leaders established or enlarged numerous scientific, technical and medical institutes and agencies in Berlin, as well as innumerable industrial and cultural projects. The Reichsanstalt was a welcome addition to their plans. Both government and benefactor wanted it in Germany's new capital.

Strictly speaking, however, the Reichsanstalt's home was not Berlin, but rather the city of Charlottenburg, which until 1920 was a self-administered political entity whose eastern border adjoined Berlin's western border and which was located about 3 kilometres due west of the Brandenburg Gate. Until about 1880 Charlottenburg remained primarily a *Residenzstadt*: influential financiers and industrialists such as Gerson von Bleichröder, Robert Warschauer and Werner Siemens; eminent academics such as Gustav Schmoller and Theodor Mommsen; leading artists such as Julius Wolff and Christian Rauch; and various high government officials — these and other prominent members of the Prussian and Reich élite had their villas there. Charlottenburg's country-like atmosphere provided them with an antidote to the daily environmental and social problems they encountered in Berlin.[3] The noise from Berlin was scarcely audible in Charlottenburg, the visitors were few, and the air was 'filled with the scent of lime trees and oranges'.[4]

In the 1880s the scent began to change. The growth of new industries in Berlin and then in Charlottenburg slowly transformed the latter's character. The middle classes began to settle in and the city experienced a population explosion: from about 31 000 residents in 1880 to about 306 000 in 1910.[5] During the Imperial period (1871–1918) Charlottenburg changed from a sparsely populated country-like town into a densely packed big industrial city.[6] None the less, in the late 1880s it still offered plenty of disturbance-free land for an institute devoted to exacting physical research and testing. Moreover, the city housed the new Technische Hochschule Charlottenburg, completed in 1884, where temporary space was promised for the Reichsanstalt's Technical Section. With the Reichsanstalt's establishment in 1887, Charlottenburg was on its way to becoming Germany's technopolis.

Siemens' circumstances dictated Charlottenburg as the choice for the Reichsanstalt's plant. Behind his estate and only a few hundred metres away from the new Technische Hochschule Charlottenburg, Siemens owned a large tract of land. In 1885 he gave this tract — a total of 19 800 square metres valued at 566 157 marks — for the Reichsanstalt.[7] More particularly, he gave it as the site for the Scientific Section. Both he and his good friend Hermann von Helmholtz, Germany's foremost man of science and future first President of the Reichsanstalt, expected that this large tract would provide enough space for the Scientific Section's present and future needs. For, even prior to the Reichsanstalt's formal approval by the government and the legislature, they foresaw the need to plan for the Reichsanstalt's future expansion. 'Once such an institute exists', Helmholtz said in 1884, 'it will become not only a workplace for its own employees, but also a place of work for those individuals who have distinguished themselves scientifically. As science progresses, the institute will have to become bigger. Therefore one should not select a site [*Entwicklungsbasis*] that is too small for future development.'[8] Moreover, both Siemens and Helmholtz hoped that either Siemens' gift of land by itself or in combination with some future acquisition of additional land would provide enough space eventually to locate the Reichsanstalt's Technical Section next to the Scientific.

That acquisition took place in 1892, shortly before Siemens' death in December of that year. In order to build a new home for the Technical Section, the Reich purchased from Siemens an additional 14 389 square metres costing 373 106 marks.[9] Siemens' desires to aid pure physics in general and his friend Helmholtz in particular had led him to offer the Reich land for the proposed Reichsanstalt if the Reich agreed to construct and equip a *scientific* building along with a residence for the Reichsanstalt's President. By 1893 the Reichsanstalt occupied 34 189 square metres (or about 8.45 acres) of land, valued at 939 263 marks. Prior to World War I, at least, the Reichsanstalt occupied the world's largest site for physical research.

That site provided the Reichsanstalt's leaders with sufficient space in which to develop the Scientific Section and, in due course, the Technical Section as well. It also provided them with security against external physical influences that might impinge on their exacting physical measurements. More precisely, it provided them with such security during the better part of the Reichsanstalt's first decade. For in 1895, shortly after Friedrich Kohlrausch succeeded Helmholtz as the new Reichsanstalt President, the Reichsanstalt found itself compelled to protect its working conditions against a new set of foreign physical disturbances. For the combined forces of the city of Charlottenburg, the firm of Siemens & Halske, and the Berlin–Charlottenburg Electrical Streetcar Company (*Berlin-Charlottenburger Strassenbahngesellschaft*) battled for the right to run electric streetcars in front of the Reichsanstalt. The battle owed its origin to the fact that electromagnetic fields emanated from uninsulated electrical cables placed at street level, passed through the earth as so-called vagabonding currents, and rendered all nearby magnetic and many electrical measurements nugatory. The Company wanted to place such cables along its tracks, which lay only 280 metres from the Reichsanstalt.[10] Moving streetcars, moreover, sent mechanical vibrations through the ground and air which likewise invalidated precision-measurement work. From the start of his presidency, therefore, the electric-streetcar problem caused Kohlrausch and his associates 'a lot of work and concern'. 'How can we claim to be the world's foremost physical laboratory', he wrote to a friend, 'if in fact . . . we are unable to conduct all finer electrical and magnetic work . . . ? That is simply unacceptable. Hopefully, we can still find a way out.'[11] The presence of electric streetcars constituted his greatest political problem with industry and local government.

In an address of 1895 to the Elektrotechnischer Verein on the disturbances to scientific institutes caused by electrical streetcar systems, Kohlrausch made it clear that his and the Reichsanstalt's battle with its local municipal and commercial interests was not unique. As to the Reichsanstalt's own situation, he noted that, despite the Reichsanstalt's natural affinities to industry and technology, it could not allow the destruction of its scientific work by the unchecked plans of industrial technology, including those of the Berlin–Charlottenburg Electrical Streetcar Company.[12] Nor could other physical institutes allow their work to be affected. For, from the mid-1890s on, the construction of electrical streetcar systems in the major cities of Western Europe and North America caused similar problems at other urban scientific institutes and affected telegraph systems as well. Physicists engaged in a 'fierce fight' with the streetcar companies over their plans to lay uninsulated tracks near physics institutes.[13] And as physical research became ever more precise, it became correspondingly more important to protect that research against foreign disturbances. The ability of urban physics institutes — above all, the Reichsanstalt's — to conduct precision measurements was at stake. 'It was a bad time my first year, or perhaps the first few years', Kohlrausch later wrote. 'Having departed Strasbourg [where Kohlrausch had previously inhabited the local professorship of physics] in conflict against the earth currents produced by the electrical streetcars, I found the situation vis-à-vis the public in Charlottenburg even more difficult.' He considered it particularly unpleasant and unfortunate that the Reichsanstalt's interests conflicted with those of Siemens & Halske as well as with the city of Charlottenburg:

The Reichsanstalt — still in the process of growing — was still foreign to the spirit of Charlottenburg. And now this new organization [i.e. the Reichsanstalt] opposed the [city's] interests. The mayor, the city council, and the townspeople felt themselves injured and insulted by the intruder. There were hard words at the town council meetings, especially since it turned out that the officials making the decisions took seriously the Reichsanstalt's objection against the presence of earth currents. Antipathy towards the Reichsanstalt's other interests also came in the form of limited cooperation from city officials. It was really a bad time, this struggle with one's own city as well as with the Electrical Streetcar Company and with Siemens & Halske.[14]

It took six years, from 1895 to 1901, for the parties to negotiate a settlement. The Reichsanstalt finally agreed to allow the Company to run its cars past the Reichsanstalt, but only after the Company had agreed to use doubly insulated overhead electrical cables. Moreover, the Company compensated the Reichsanstalt with 100 000 marks to be used towards the future construction of a disturbance-free magnetic laboratory.[15] Shortly thereafter, it is worth noting, Kohlrausch and others managed to develop a disturbance-free (i.e. astatic) torsion magnetometer; they replaced the magnetometer needle by an astatic system that made the fields generated by the streetcars inconsequential.[16] With an astatic magnetometer, as well as with a so-called *Panzergalvanometer* developed by Henri Du Bois and Heinrich Rubens, the Reichsanstalt could conduct its precision electrical and magnetic work without fear of external disturbances. By 1902 the first electric streetcars were running past the Reichsanstalt. The Reichsanstalt had found a *modus vivendi* with its political and industrial environment.

2 Architectural Design: The Reichsanstalt's Plant

Between 1887 and 1896 the Reichsanstalt's physical plant slowly took shape. It consisted of ten buildings, five for each section. The buildings were kept as far apart from one another as possible, so as to minimize any potential disturbances emanating from the various buildings and so as to maximize the amount of sunlight reaching each building. The land not occupied by these buildings constituted a finely landscaped garden which served both aesthetic and physical ends. Except for the establishment of a small thermometer-testing station in Ilmenau in 1889 and a disturbance-free magnetic laboratory near Potsdam in 1913, all of the Reichsanstalt's facilities were located in Charlottenburg. (See Figure 8.1.)

Two professional state architects helped realize the general plans of Siemens and Helmholtz, who themselves contributed to the design of the Scientific Section, and fashioned the overall layout of the Reichsanstalt's plant. In 1884 and 1885 Paul Spieker, a Prussian civil servant, who had previously planned and overseen construction of the University of Berlin's Physical Institute as well as the Astrophysical Observatory in Potsdam, designed the buildings of the Scientific Section. Thereafter, Theodor Astfalck, an architect from the Reich Ministry of the Interior, who, like Spieker, had previously been involved in constructing the Observatory in Potsdam, completed the architectural details left unfinished by Spieker and supervised the Reichsanstalt's construction.[17] To realize their ideas, Siemens and Helmholtz chose men experienced in the design and construction of the most modern scientific buildings. As was the case with academic physics buildings, the construction of the Scientific Section's buildings was a drawn-out affair.[18] It began only after the legislature finally approved the Reichsanstalt in March 1887 and ended in 1891, some four years later.

The Scientific Section's main building, known as the *Observatorium*, was

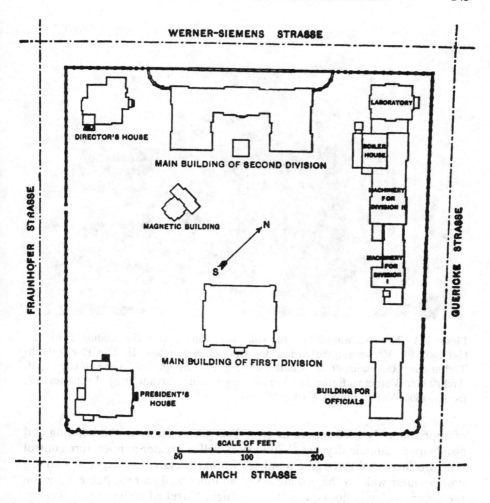

Figure 8.1 Layout of the Reichsanstalt. From Carhart (1900), p. 565.

the Reichsanstalt's central and primary structure (Figure 8.2). The location and architectural style of all other Reichsanstalt buildings were dependent on the decisions made about the location and architecture of the *Observatorium*. Above all, the building's planners sought to achieve two structural characteristics: maximum amount of freedom from outside disturbances, especially those due to mechanical vibrations and alien electromagnetic phenomena; and maintenance of temperature regularity and control.[19]

To achieve a disturbance-free structure, the Reichsanstalt's planners made a number of careful decisions and spent a good deal of money. First, they located the building towards the centre of the grounds — that is, as far away from street traffic and the other buildings as conditions would permit.[20] Second, they built the entire structure upon a two-metre-thick concrete slab

Figure 8.2 The Reichsanstalt's *Observatorium* or main scientific building. From 'IV Gebäude für Verwaltungsbehörden des Deutschen Reiches. II. Die Physikalsich-Technische Reichsanstalt in Charlottenburg', in *Berlin und seine Bauten* (ed. Architekten-Verein zu Berlin und Vereinigung Berliner Architekteu), 3 volumes in 2, Berlin, 1896, Vol. 2, pp. 80–4, on p. 81.

of approximately 1000 square metres. This gave solidity to the building and also helped maintain dryness.[21] Solidity as well as greater temperature control was also increased through a number of other features: an especially strong double outer wall on the ground floor; an inner wall and vaulted ceilings on the other two main floors; a half-metre-high, barrelled-arch crown covering; reinforced curved girths; and terrazzo floors[22] (Figure 8.3).

To achieve temperature regularity or the greatest degree possible of temperature control, especially for the interior workrooms, the planners oriented the axis of the nearly square building due north. Because most laboratory work was done in the morning, the lack of sunlight on two sides of the building helped to maintain the building's temperature control.[23] Steam heating, electric lighting and a built-in air-ventilation system throughout the entire building helped keep the building's temperature at 20 °C.

The *Observatorium* consisted of a basement, three main floors and a dome.[24] Resting on the two-metre-thick concrete foundation, the basement served principally as a means of isolating the building's heat from ground dampness. It was also occasionally used for observational work requiring a high degree of stability. The three main floors were all of similar layout and construction. The centre and largest room of each floor admitted sunlight

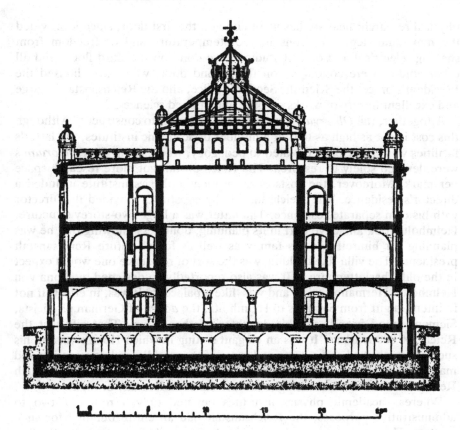

Figure 8.3   Cross-section of the Reichsanstalt's *Observatorium*. From 'IV Gebäude für die Verwaltungsbehörden des Deutschen Reiches. II. Die Physikalsich-Technische Reichsanstalt in Charlottenburg', in *Berlin und seine Bauten* (ed. Architekten-Verein zu Berlin und Vereinigung Berliner Architekteu), 3 volumes in 2, Berlin, 1896, Vol. 2, pp. 80–4, on p. 83.

only through a skylight, the amount of admitted sunlight naturally being greatest on the top floor and correspondingly less on the middle and lower floors. This, too, helped control the building's temperature. So, too, did the thick walls that enclosed these rooms and the four copper ovens, with thermostats, that controlled the temperature to within several one-hundredths of a degree. Moreover, each of these three centre rooms had three of their four walls surrounded by a horseshoe-shaped corridor that further limited outside disturbances due to mechanical vibrations or temperature deviation. These rooms, where the Reichsanstalt's main scientific work was performed, were excellent workplaces for physical and precision-technical work. Around them were the remaining rooms for experimental work and administration.

The building's individual floors were devoted to specific branches of physical research: heat studies were done on the first floor, since it provided the maximum degree of constancy of temperature and of freedom from shaking; electrical and optical studies were done on the third floor; and all other studies were conducted on the second floor, which also housed the President's office, the Scientific Section's office, and the Reichsanstalt's 'large and excellent library of works on pure and applied science'.[25]

Altogether, the *Observatorium* cost 387 000 marks to construct.[26] Although this cost is not as high as that of a number of academic institutes, the latter's facilities were principally devoted to pedagogy, whereas the *Observatorium*'s were devoted solely to research. The Reichsanstalt got more research space per mark. Moreover, the costs of an academic physics institute included a director's residence. The Reichsanstalt, by constrast, provided its Director with his own separate residence. The latter was a large two-storey structure. Helmholtz took an active part in its planning, doubtless knowing that he was planning for himself and his family as well as for all future Reichsanstalt presidents. The villa-like building was the sort of residence one would expect in the old Charlottenburg.[27] It was also reportedly co-designed by Anna von Helmholtz, Hermann's wife, and was 'like a palace'.[28] It was, in effect, if not in intent, a gift from Siemens to Helmholtz, the *doyen* of German physicists, Siemens' good friend, and his daughter's father-in-law. That gift cost the Reich 100 000 marks.[29] It was an elegant setting in which Helmholtz and his successors received and entertained some of the world's leading scientists and many of Germany's social, industrial and political élite — and in which Reichsanstalt subordinates might contact the President day or night.

Whereas academic physics institutes devoted only a room or two to administrative purposes, the Reichsanstalt had an entire building for such matters. The very existence of an administration building — probably the first of its kind for a physical institution — suggests the presence of a bureaucratic aspect to the Reichsanstalt. This two-storey building cost 100 000 marks to construct. It housed the Scientific Section's business office, the conference room of the Reichsanstalt's Board of Directors and the Board President's office, and provided living quarters for minor Reichsanstalt officials.[30]

To help maintain the *Observatorium*'s goals of freedom from outside disturbances and temperature control, the Reichsanstalt's planners established a separate, small, iron-free magnetic house and a separate machine house. The latter contained the Reichsanstalt's boiler plant; a Deutzer twin gas motor for operating storage batteries; and a direct-current dynamo. In addition to housing observation rooms and a chemical laboratory, the machine house also contained a low-temperature room for conducting scientific studies involving temperatures below freezing. The low-temperature room was surrounded by thick walls, buttressed by arches, supported by a one-and-a-half-metre-thick cement floor and illuminated solely by four skylights. Supplied with a Lindean 'ice machine' for producing liquid air and a refrigerator for producing ice and for cooling the observation room, the Reichsanstalt's workers were able quickly to reach and maintain the low temperatures needed for their scientific work. The Scientific Section's machine house was located some 60 metres from the *Observatorium* and cost 50 000 marks to construct. (The small magnetic house cost 8500 marks.) The availability of such funds, and the care and foresight with which the machine house and the other secondary buildings of the Scientific Section were planned and constructed, allowed the *Observatorium* to be supplied with a sufficient amount of electrical, magnetic and thermal power without thereby adversely affecting the building's goals of freedom from outside disturbances and temperature control.[31]

Construction of the Scientific Section's five buildings — the *Observatorium*, the President's residence, the administration building, the magnetic house and the machine house — cost a total of 644 754 marks. A further 232 000 marks was allowed for paving, landscaping, a water and drainage system, heating, water, gas and electrical connections, furniture, and so on.[32] In addition, the Reich reportedly spent another 82 310 marks on (initial) equipment and machinery for the Scientific Section.[33] Thus, the Scientific Section cost the Reich a total of 959 064 marks to build and equip (Table 8.1). Except for the Berlin and Leipzig Physical Institutes, the Reichsanstalt's Scientific Section was itself the most costly physical institute in Imperial Germany.[34] Moreover, and as noted above, those and other academic institutes were basically built for pedagogical purposes, whereas the Scientific Section was built for physical research. For the same reason, the Reichsanstalt probably provided more floor space and better facilities for research than any German academic physics institute.

The establishment of the Technical Section's laboratory buildings required some six years of planning and negotiating (1889–1894)[35] and then an additional three years of construction time. Only in 1897, a decade after it had officially begun its operations, did the Technical Section fully occupy its own buildings. The postponed construction of its facilities was due to three related sets of political, financial and ideological reasons.

First, Siemens and others wanted to use the rooms at the Technische Hochschule in order to achieve some immediate results for German industry. Scepticism voiced by some legislative and industrial opponents during Siemens' fight to establish the Reichsanstalt had still to be overcome.[36] Siemens was anxious to demonstrate to German industry and the Reichstag the benefits they might derive from the Reichsanstalt. He was opposed to constructing new buildings for the Technical Section, because he thought that would delay the immediate operation of the Technical Section.[37]

Second, the postponed construction was also due to the government's desire to avoid spending too much money on the Reichsanstalt. 'In the interests of the Reich's present financial situation', an official of the Ministry of the Interior asked Siemens and other founders of the Reichsanstalt in 1884 that 'the Reich's sacrifice not be made too large and that the new construction be limited to the physical institute [i.e. the Scientific Section]'.[38]

Third, and finally, Siemens' and Helmholtz's ideology concerning the

Table 8.1 Costs of establishing the Reichsanstalt, 1887–97

		Marks
I.	*Scientific Section*	
	Observatorium	387 000
	Machine house	50 000
	Administration building	100 000
	President's residence	99 254
	Small magnetic house	8 500
	Paving, landscaping, water and drainage system, supervision of construction, etc.	174 000
	Heating, water, gas, and electrical connections (*'innere Einrichtungen'*), furniture, etc.	58 000
	Initial equipment and machinery	82 310
	Subtotal	959 064
II.	*Technical Section*	
	Acquisition of land	373 106
	Main building	922 000
	Laboratory building	218 000
	Machine (and boiler) house	180 000
	Director's residence	140 000
	Additional plant costs	348 000
	Fittings and furniture	108 300
	Initial equipment and machinery	471 390
	(Less reduction for the year 1895/6)	(47 500)
	Subtotal	2 713 296
	TOTAL	3 672 360

relations of science and technology led to the postponement of the Technical Section. For both men regarded science as the basis of technology.[39] That standpoint naturally led them to support the construction and equipping of the Scientific Section before that of the Technical. It was a standpoint that fitted in well with the needs of German industry and the financial limitations of the German government.

Between 1893 and 1896, however, the Reichsanstalt's leaders constructed an entirely new set of five buildings for the Technical Section: the laboratory building, a machine (and boiler) house, the main or technical building, a small ventilator house and a residence for the Director of the Technical Section.[40] Astfalck, in consultation first with Helmholtz and then with Helmholtz's successor Kohlrausch, designed and oversaw the construction of these buildings, just as he had done for the Scientific Section. As one might expect, these buildings were architecturally similar to those of the Scientific Section.[41]

The Technical Section's buildings were located on the newly purchased plot of land adjacent to the north-western border of the Scientific Section's site. The determining factor in arranging the Technical Section's buildings was the necessity of placing the laboratory building, which housed the chemical laboratory on its top floor, at the extreme northern tip of the Reichsanstalt's entire site, so that the noxious fumes produced in the chemical laboratory could be blown away by the prevailing west winds.[42] The laboratory building's basement contained rooms for the high-voltage electrical laboratory, the latter's workshop, a photographic darkroom, a storage and chemical room, and two rooms for experiments in radiotelegraphy. The first floor housed rooms for studies on direct current, electrical meters and the testing of electrical apparatus; the second was reserved for investigations on self-induction and capacity.[43] The building cost 218 000 marks to erect.[44]

The laboratory building was connected to the Technical Section's machine (and boiler) house, which itself was attached to the Scientific Section's nearly identical machine house. The Technical Section's machine house contained three large boilers for heating the Section's buildings and powering the high-voltage laboratory's thirty-five-horsepower generator. A machine room contained machine-testing apparatus — in particular, load resistances, direct- and indirect-current machines and motors, and an electrical switching panel. There was also space here for making indirect-current investigations and studying transformers. Next to the machine room were the alternating-current and high-voltage electrical rooms, equipped with an 11 000-volt battery and a 40 000-volt single-phase transformer. Above the machine room was the Reichsanstalt's mechanical workshop. Finally, there was a plumber's workshop, a forge and a carpenter's workshop here.[45] The machine house cost 180 000 marks to erect.[46]

The Technical Section's third structure was the main or technical building, located directly across from the Scientific Section's *Observatorium* (Figure 8.4). This four-storey, U-shaped building cost 922 000 marks and took two years to erect.[47] Only its main walls and floors were laid down permanently; individual workrooms could be easily and variously shaped by inserting or removing partitions.[48]

Each of the main building's floors was principally devoted to one individual field of study: optics, precision mechanics, electricity and magnetism, and heat and pressure. The ground floor was divided into two halves. One half contained various offices, a reference library, and the Director's laboratory and office; the other housed the optical laboratory, with its extensive area for photometric work. Heliostats were placed in the windows of the optical (and several other) laboratories. The basement housed residences for the steward and the machinist, and contained the Section's precision-mechanical labora-tory. The latter's rooms were particularly appropriate for work demanding constancy of temperature and for studies on graduated scales (straight and circular). Moreover, below the mechanical laboratory was a one-metre-thick concrete plate used for studies requiring a high degree of freedom from vibration.

Like the ground floor, the first floor was also divided into two halves: one half contained the low-voltage electrical laboratory; the other, the magnetic laboratory. The second floor contained the Section's heat and pressure

Figure 8.4 The main building of the Technical Section. From Physikalisch-Technische Bundesanstalt, Braunschweig.

laboratory, along with a large conference room. Finally, a small domed area in the centre of the building's roof provided extra space for optical experiments, a photographic darkroom, and a storage room for old or currently unused apparatus and materials.

The Technical Section also provided a small ventilator house and a home for its Director. The latter not only housed the Section's Director, but also contained rooms for two assistants and a house porter.[49] It cost 140 000 marks to construct.[50] Ernst Hagen, the Director of the Technical Section from 1893 to 1918, claimed that the availability of this official residence was crucial for his own scientific work. Such a residence, he told Heinrich Kayser in 1911, 'is the prerequisite for a [?] physicist who has to direct a laboratory and who also wants to be able to get to his own work. Do you really believe', he asked Kayser, 'that during my time here I could have done even a single one of the numerous studies with [Heinrich] Rubens if I had not had an official residence?'[51]

In addition to the previously mentioned land costs (373 106 marks) and building construction costs (1 460 000 marks), the Reich spent an additional 880 190 marks on the Technical Section, with 348 000 marks going for plant costs (street paving, a sewage system, gas and heating equipment, construction design, and so on); 108 300 marks for fittings and furniture; and 471 390 marks for initial equipment and machinery. By the fiscal year 1897/8, the

Reich had provided 2 713 296 marks to purchase land for, erect and equip the Technical Section.[52] (See Table 8.1.)

As Table 8.1 shows, the Reich spent nearly three times as much money on the Technical Section as on the Scientific Section. Whether the significantly higher costs of the Technical Section should be interpreted as a sign of the Reichsanstalt's greater interest in technology, as opposed to science, or whether it should be simply ascribed to a presumably greater cost in conducting technological research and testing, as opposed to scientific research, is uncertain. What is certain is that the 3 672 360 marks spent to establish the Reichsanstalt's physical plant was more than double the 1 500 000 marks (equivalent) spent by the United States to construct and equip its National Bureau of Standards by 1903 and more than six times the 600 000 marks (equivalent) spent by Britain to construct and equip its National Physical Laboratory by 1902.[53] By 1897, a full decade after Siemens had established the Reichsanstalt, Kohlrausch rightly claimed that the Reichsanstalt 'had now secured the most important basic conditions for pursuing its tasks . . . in a rich and appropriate style'. 'And it is fair to say', he boasted, 'that the noble establishment that has been produced is matched by no other in the world.'[54] His boast was not merely Germanic chauvinism and pride. A contemporary British observer wrote: 'The question as to whether that institution is not too magnificent has in fact occurred to many of those who have seen it.'[55]

If the minor issue of the Reichsanstalt's magnificence was open to debate, the major one concerning its environmental control was not. For, as we have seen, between 1887 and 1901 its leaders had managed to secure an internally and externally controlled environment. The completed Reichsanstalt laboratories represented an unmistakable sign of Germany's keen interest in advancing physical metrology, precision technology and testing. By the turn of the century, the Reichsanstalt had the largest and foremost physical plant in these fields; it probably remained so until after World War I. Both substantively and symbolically, the Reichsanstalt was a product of the new Imperial Germany. Its laboratories became the models for others.

Notes and References

1. Cahan (1982). For a full discussion of the Reichsanstalt in Imperial Germany, see Cahan (forthcoming).
2. Cahan (1982).
3. Denke (1956), p. 108; Haemmerling (1955), p. 113; Leyden (1971), pp. 322–4.
4. Hertz to parents, 22 June 1880, in Hertz (1927), p. 97.
5. Leyden (1933), p. 207.
6. Haemmerling (1955), p. 18; Leyden (1971), p. 324.
7. 'Taxé des vom dem Kaiserlichen Geheimen Regierungsrathes Herrn Dr. Siemens dem deutschen Reiches übereigneten Grundstückes', 16 November 1885, signed Bohl, Königlicher Baurath, in 'Kommission zur Vorberathung des Planes für eine physikalisch-technische Reichsanstalt', June 1883–November 1885, Zentrales

Staatsarchiv, Potsdam, Reichsamt des Innern, Nr. 13144/8, Bl. 259–62; hereafter cited as ZStA, Potsdam, RI.

8. 'III. Sitzung der Subkommission', Berlin, 29 October 1884, Siemens Museum, Munich, Siemens Archiv Akten 61/Lc 973; hereafter cited as SAA. See also Kohlrausch's report of 9 August 1887 and Board Session 4, 25 March 1889, *ibid.*

9. 'Anlage IV. Reichsamt des Innern', in *Haushalts-Etat des Deutschen Reichs: Reichshaushalts-Etat für das Etatsjahr 1892/93 nebst Anlagen* [Berlin, 1892], pp. 40–1.

10. 'Vermischtes', *Centralblatt der Bauverwaltung* (1895), **15**, 276. Junk (1905) reported (pp. 205–6) that the side-effects of electric streetcars could produce measuring errors from 100 to 140 times beyond normal.

11. Kohlrausch to H. Wagner, 8 May 1895, Niedersächsische Staats- und Universitätsbibliothek Göttingen, Handschriftenabteilung, Sig. H. Wagner 29.

12. Kohlrausch (1895), reprinted in Kohlrausch (1910–11), Vol. 1, pp. 886–901. For the electrotechnologists' own concerns about and investigations of this problem see Naglo (1904), pp. 58–9 and 76–9; *Der Verband Deutscher Elektrotechniker 1893–1918*, n.p., 1918?, pp. 41–2; and especially Görges (1929), pp. 262–70 (and the references therein).

13. Holborn (1895), p. 180. See also the relevant letters in ZStA, Potsdam, RI, Nr. 13144/3; and, on the disturbances by the new electric streetcars at the physics institutes at Halle, Darmstadt and elsewhere, Junk (1905), pp. 205–6.

14. Kohlrausch, quoted in Heydweiller (1910–11), p. lvii.

15. 'Besprechung im Reichsamt des Innern am 2. November 1901 betreffend die Anlage eines von magnetischen Störungen freien Hülfslaboratoriums der Physikalisch-Technischen Reichsanstalt', ZStA, Potsdam, RI, Nr. 13144/4, Bl. 148–51. For details of the agreement between the Reichsanstalt and the Berlin-Charlottenburg Streetcar Company, see Kohlrausch to Staatssekretär, Reichsamt des Innern, 10 February 1902, Bundesarchiv Koblenz: R2 12376. The entire battle can be followed in greater detail through the materials in ZStA, Potsdam, RI, Nr. 13144/3. Cf. the disturbances to precision-measurement work at the Physics Institute of the Technische Hochschule Karlsruhe, where, after long negotiations, the local streetcar company compensated the Institute with 60 000 marks (Lehmann, 1911, pp. 80–1). On the similar problems at Munich, see Wien (1926), p. 209; and at Jena, which was forced to construct another building away from the streetcars, see Auerbach (1922), pp. 113–14.

16. 'Die Tätigkeit der Physikalisch-Technischen Reichsanstalt im Jahre 1902', *Zeitschrift für Instrumentenkunde* (1903), **23**, 113–25, 150–7 and 171–4, on pp. 121; and 'Die Tätigkeit der Physikalisch-Technischen Reichsanstalt im Jahre 1903', *ibid.* (1904), **24**, 133–47 and 167–80, on pp. 138–9.

17. See Spieker to Siemens, 17 February and 20 February 1884; Gossler to Siemens, 29 December 1884; Helmholtz to Spieker, 28 March 1885; 'VIII. Sitzung der Subkommission, Berlin, 1. Mai 1885'; and 'IX. Sitzung der Subkommission, Berlin, 13. Juli 1885'; all in SAA 61/Lc 973. Helmholtz to ?, 12 January 1885'; and (two letters of) Helmholtz to Spieker, 28 March 1885, copies in 'Kommission zur Vorberathung', ZStA, Potsdam, RI, Nr. 13144/8, Bl. 150–1, 161–2 and 178, respectively; and Siemens to Boetticher, 21 July 1886, ZStA, Potsdam, RI, Nr. 13144/2, Bl. 4–6. On Spieker, see 'Dr. Paul Spieker', *Zentralblatt der Bauverwaltung* (1896), **6**, 541–2; Junk (1905), p. 143; and 'Das physikalische Institut', in *Berlin und seine Bauten* (ed. Architekten-Verein zu Berlin und Vereinigung Berliner Architekten), 3 vols. in 2, Berlin, 1896, Vol. 2, pp. 264–9, on p. 268. On Astfalck, see Weymann to Maybeth, 3 August 1886, ZStA, Potsdam, RI, Nr. 13144/2, Bl. 8–10; and B., 'Theodor Astfalck', *Zentralblatt der Bauverwaltung* (1910), **30**, 88.

18. Cahan (1985), pp. 18–19.

19. 'Denkschrift betreffend die Errichtung einer "physikalisch-technischen Reichsanstalt" für die experimentelle Förderung der exakten Naturforschung und der Präzisionstechnik', in Germany, Reichstag, *Verhandlungen. B. Sammlung sämtlicher Drucksachen des Reichstages*, Berlin, 1887, Drucksache Nr. IV, Beilage B, pp. 37–50, on pp. 45–6 (hereafter cited as 'Denkschrift, 1887'); and 'IV. Gebäude für die Verwaltungsbehörden des Deutschen Reiches. 11. Die Physikalisch-Technische Reichsanstalt in Charlottenburg', in *Berlin und seine Bauteng* (ed. Architekten-Verein zu Berlin und Vereinigung BerlinerArchitekten), 3 vols. in 2, Berlin, 1896, Vol. 2, pp. 80–4, on pp. 82–3.

20. 'IV. Gebäude für die Verwaltungsbehörden', *op. cit.* [Note 19], Vol. 2, p. 81.

21. 'Denkschrift, 1887', [*op. cit.*, Note 19], p. 46.

22. Spieker (1888), pp. 564–6; Pernet (1891), p. 2; and 'IV. Gebäude für die Verwaltungsbehörden', *op. cit.* [Note 19], Vol. 2, p. 83.

23. 'Denkschrift, 1887', *op. cit.* [Note 19], p. 45.

24. This description of the *Observatorium* follows that presented in Hagen and Scheel (1906), pp. 60–5; Pernet (1891), pp. 2–3; and 'IV. Gebäude für die Verwaltungsbehörden', *op. cit.* [Note 19], Vol. 2, pp. 82–4.

25. Galton (1895), quote on p. 606.

26. 'Anlage IV. Reichsamt des Innern', in *Haushalts-Etat des Deutschen Reichs: Reichshaushalts-Etat für das Etatsjahr 1887/88 nebst Anlagen* [Berlin, 1897], p. 33.

27. 'IV. Gebäude für die Verwaltungsbehörden', *op. cit.* [Note 19], Vol. 2, p. 81.

28. Extract from Otto Warburg's taped autobiography, quoted in Hans Krebs (1979), p. 94.

29. 'Anlage IV. Reichsamt des Innern', *op. cit.* [Note 26], p. 33.

30. *Ibid.*, p. 33; 'IV. Gebäude für die Verwaltungsbehörden', *op. cit.* [Note 19], Vol. 2, pp. 81–2.

31. 'Anlage IV. Reichsamt des Innern', *op. cit.* [Note 26], p. 33; Pernet (1891), p. 3; 'IV. Gebäude für die Verwaltungsbehörden', *op. cit.* [Note 19], Vol. ii, p. 82; Hagen and Scheel (1906), pp. 61–2; and 'Anlage IV. Reichsamt des Innern', *Haushalts-Etat des Deutschen Reichs: Reichshaushalts-Etat für das Etatsjahr 1891/92 nebst Anlagen*, Berlin, 1891, p. 36.

32. 'Anlage IV. Reichsamt des Innern', *op. cit.* [Note 26], p. 33.

33. Carhart (1900), p. 562.

34. Cahan (1985), p. 17.

35. Board Session 4, 25 March 1889; Board Session 2, 16 March 1891; Staatssekretär to Siemens, 10 October 1891; all in SAA 61/Lc 973. 'IV. Reichsamt des Innern [for Etatsjahr 1892/93]', p. 41; and 'IV. Reichsamt des Innern', *Haushalts-Etat des Deutschen Reichs: Reichshaushalts-Etat für das Etatsjahr 1893/94 nebst Anlagen*, Berlin, 1893, pp. 40–1.

36. Cahan (1982), pp. 278–83.

37. 'IV. Sitzung der Subkommission, Berlin, den 3.11.1884', ZStA, Potsdam, RI, Nr. 13144/9, Bl. 69–74; and 'VIII. Sitzung der Subkommission, Berlin, 1. Mai 1885', SAA 61/Lc 973.

38. 'Erste Sitzung der Kommission zur Berathung der Organisation und des Kostenanschlags für das mechanisch-physikalische Institut, abgehalten im Reichsamt des Innern, Mittwoch den 15. Oktober 1884 Morgens 10 Uhr', SAA 61/Lc 973.

39. Cahan (1982), pp. 263–8.

40. Hagen and Scheel (1906), p. 62.

41. 'IV. Gebäude für die Verwaltungsbehörden', *op. cit.* [Note 19], Vol. 2, p. 84.

42. *Ibid.*

43. Hagen and Scheel (1906), p. 65.

44. 'Anlage IV. Reichsamt des Innern, [for Etatsjahr 1893/94]', *op. cit.* [Note 35], p. 41.

45. Hagen and Scheel (1906), pp. 64–5.

46. 'Anlage IV. Reichsamt des Innern [for Etatsjahr 1893/94]', *op. cit.* [Note 35], p. 41.

47. *Ibid.*; and 'Denkschrift über die Tätigkeit der Physikalisch-Technischen Reichsanstalt von Frühjahr 1895 bis zum Sommer 1897', *Sammlung sämmtlicher Drucksachen der Reichstages*, 9th Legislatur-Periode, 5th Session, 1897/98, Aktenstück Nr. 80, pp. 870–83, on p. 870.

48. Hagen and Scheel (1906), p. 62. This description of the Technical Section's main building relies exclusively on *ibid.*, pp. 62–4.

49. *Ibid.*, p. 65.

50. 'Anlage IV. Reichsamt des Innern [for Etatsjahr 1893/94]', *op. cit.* [Note 35], p. 41.

51. Hagen to Kayser, 25 May 1911, Staatsbibliothek Preussischer Kulturbesitz, Berlin, Slg. Darmstaedter, F i c 1902 (6), Bessel-Hagen.

52. This grand total is based on the figures given in 'Anlage IV. Reichsamt des Innern', *Haushalts-Etat des Deutschen Reichs . . .* for the years 1893 to 1898, along with those provided by Carhart (1900), pp. 561–2.

53. Forman *et al.* (1975), pp. 92 and 95.

54. 'Denkschrift . . . vom Frühjahr 1895 bis zum Sommer 1897', *op. cit.* [Note 47], p. 871.

55. 'The National Physical Laboratory', *Nature, Lond.* (1898), **58**, 565–66, p. 565.

9

Building England's First Technical College: The Laboratories of Finsbury Technical College, 1878–1926

W. H. Brock

Like the 1960s, the 1870s were a golden age for new academic buildings, not merely in Great Britain, but also in Europe generally. Between 1873 and 1874 the Cavendish Physical Laboratory was opened at the University of Cambridge to the designs of W. M. Fawcett, with fittings devised by James Clerk Maxwell based upon the experiences of the Clarendon Laboratory at Oxford and of William Thomson's laboratory at Glasgow. In 1874 laboratories were opened at the new Royal Naval College at Greenwich, science colleges were opened at Bristol and Leeds, and George Cary Foster began to teach practical physics at University College London — to be followed three years later by W. G. Adams at rival Kings College. In 1878 Alfred Waterhouse's Gothic University College opened in Liverpool, with its extraordinary tiered chemical laboratory designed by James Campbell Brown; Sheffield responded with Firth College in 1879; and the decade ended in 1880 with the opening of Mason's College in Birmingham and the beginning of the new City & Guilds of London Institute's ambitious building programme for technical education — completed in 1883 as Finsbury Technical College and in 1885, at South Kensington, with Waterhouse's Queen Anne-style Central Institution.

The best source of information on this frenzied British activity (for which there is no comparable continental equivalent) is the large illustrated volume *Technical School and College Buildings*, by the English architect Edward Cookworthy Robins (1830–1918).[1] Although other professional architects, such as Alfred Waterhouse and Aston Webb, designed many more prestigious academic buildings than Robins, he seems to have given much more detailed thought to the subject of designing buildings and interiors for scientific and technological education and research than his peers. Waterhouse's record is certainly impressive in quantity and in its geographical distribution: Gonville & Caius College, Cambridge (1868–71); Owens College (1870–3); the Yorkshire College at Leeds (1878); Liverpool's University College (1878); the City & Guilds College, South Kensington (1881–5); Girton College, Cambridge (1873–81); and the Natural History Museum, South Kensington (1873–81).[2] However, Waterhouse put up precisely what he was told to do and dressed the buildings as his clients could afford. That his science buildings were effective owed more to the quality of the technical advice he received from the scientists who were to use his buildings: H. E. Roscoe at Manchester, Thorpe at Leeds, W. E. Ayrton and Unwin at the Central Institution, Campbell Brown at Liverpool and Richard Owen at the Natural History Museum.

Robins, on the other hand, who was a close friend of the three Finsbury College 'Mohicans', Henry Edward Armstrong, William E. Ayrton and John Perry, reflected personally and deeply on the ideal environment for technical education. Moreover, as an executive member of the City & Guilds of London Institute, he was able to play an influential role in formulating its policy on technical education in the 1870s and 1880s. His book on technical school design continued to be cited by architects until the 1920s; only then was it replaced as the Bible of laboratory design by the works of Russell, Clay and Munby.[3]

Robins had begun to practise as an independent architect in 1851, after training with Sancton Wood (1815–86).[4] Like most Victorian architects in the second half of the nineteenth century, his bread and butter came from religion and education, as the churches expanded their sittings in the wake of the ominous 1851 Religious Census and as the middle classes determined to see their children fitted educationally for an industrial and commercial age. Robins designed dozens of London churches: St. Jude's and St. Saviour's at Brixton; St. John's and St. Saviour's at Wandsworth (all Anglican); and non-conformist churches at Wandsworth, Dulwich, Camden Town, and the huge tabernacle at Newington Butts for the popular Baptist preacher, Charles Spurgeon, in 1862. He also designed the Foreign Missionaries' College at Sevenoaks in 1882 and there were several contracts for churches in Madagascar. Similarly, he designed dozens of elementary schools all over London: the new Bedford Grammar School buildings (and laboratories) in 1882; the North London Collegiate School for Girls in Camden Town in 1880 (which was converted from a huge emporium); and, his most prestigious and architecturally distinctive project, the Merchant Venturers' Technical School in Bristol in 1882, where he was free to develop his influential 'hall passage system',[5] and to design laboratories on a grand scale.

Apart from the flamboyantly Gothic Merchant Venturers' School, none of these commissions was for a specifically scientific or technical institution. What, then, gave Robins the authority to deliver addresses on technical education and the special problems of designing buildings for this specialized kind of secondary education? The answer is to be found in another central concern of Victorian Britain, public health and sanitary engineering, for a large proportion of the special technical problems of designing laboratories and workshops (including factories for commercial purposes) were variations on the twin subjects central to the concern of public health experts: drains and ventilation. It is no accident that Robins was a founder member of the Sanitary Institute of Great Britain in 1877.

During the London cholera outbreak of 1852–3, Robins was inspired by George Godwin's editorial articles in *The Builder* to take an interest in health and sanitation.[6] He played a leading role in the voluntarist District Board of Health which had been set up in the Regent Square district of the parish of St. Pancras. His prominence here led to an appointment as architect to the Croydon Burial Board, for whom he designed Croydon Cemetery, one of the many new municipal cemeteries laid out to allay fears concerning the suspicious link between city burial grounds and disease which had been sensationally publicized by Edwin Chadwick, G. A. Walker and Charles Dickens.[7] His views on sources of infection (miasmatic, as befitted the period) were publicized in a pamphlet, *A Practical View of the Sanitary Question* (1854), and he joined Chadwick's Metropolitan Health of Towns Association, whose propaganda purpose was 'to diffuse among the people the valuable information elicited by recent inquiries, and the advancement of science, as to the physical and moral evils that result from the present defective sewerage, drainage, supply of water, air and light, and the construction of dwelling houses'.[8]

In 1875 Robins was appointed surveyor to the Worshipful Company of Dyers, one of the ancient Guilds of the City of London, and he was its Prime Warden in the crucial year, 1879, when the Guilds, prompted by T. H. Huxley, Prime Minister W. E. Gladstone and their own consciences, began to invest in technical education.[9] Robins was a member of the important sub-committee C (Buildings) of the City & Guilds, together with Richard Wormell, Sir Frederick Bramwell, Sir Sydney Waterlow and Sir John Watney. Robins's unpublished correspondence with the chemist H. E. Armstrong shows him to have been a zealous advocate of technical education. Such an education programme was necessary, Robins believed, not merely for manufacturers and artisans, but also for professional groups such as the architectural community to which he belonged.[10] Technical education, he wrote:

> . . . is the complement and crown of all utilitarian education as contrasted with literary culture only — it is pure science carried to its legitimate issue: that is to say, its varied applications to human requirements; it is the practical applications of scientific principles to special objects and purposes, and it is as necessary to professional men as it is acknowledged to be to manufacturers and artisans.[11]

The City & Guilds of London Institute was formed in November 1878 in response to the opinion of successive Royal Commissions on education, industrialists, leading Guildsmen and scientists that the maintenance of British industrial supremacy in the new era of electricity and synthetic chemicals depended upon the active promotion of technical education. On the advice of the Society of Arts and a number of educational experts, the 'City & Guilds' (as it became popularly known) decided to aid the cause of technical education by endowing a university-level teaching and research engineering institution within the City of London. However, because of financial difficulties (the Drapers' Company pulled out following a disagreement over policy), and the unavailability of suitable land within the city, this 'Central Institution',[12] which was finally completed in 1885, was built at South Kensington, where, as the City & Guilds College, it now forms part of the Imperial College of Science and Technology. (Waterhouse's elegant building was demolished by the College in 1962 and 'replaced by buildings designed by architects with different ideals'.[13])

Meanwhile, so as not to appear dilatory while this plan matured, the City & Guilds decided to sponsor a 'Trades School' of evening classes in the large Cowper Street Middle Class School, whose headmaster, Richard Wormell (1838–1914), had long been a friend of technical instruction.[14] Wormell agreed to lease the school premises to the City & Guilds after 4.00 p.m. and also agreed to give up space to allow a new building on part of the school's playground, despite 'a pecuniary loss to himself'.[15] And so it came about, on 1 November 1879, that a large audience of city dignitaries and local artisans came to the school's assembly hall to listen to an address on 'The improvement science can effect in our trades and the condition of our workmen'. The speaker, William E. Ayrton (1847–1908), a pioneer of electrical engineering, was inaugurating the series of evening lectures on 'The practical applications of electricity and magnetism (electric bells, light, telegraph, electric fire and burglar alarms)' which were to be held in the school's basement laboratory. Complementing these, in the school's attic, Henry Armstrong (1848–1937), the German-trained organic chemist who had already gained a local reputation for his industrial chemistry lectures at the London Institution in Finsbury Square, began to lecture and supervise laboratory work on 'The first principles of chemistry'.

Ayrton, who had designed and worked in a palatial physics laboratory at the Royal Engineering College in Tokyo as Professor of Natural Philosophy and Telegraph Engineering (1873–8),[16] and Armstrong, who had worked in Kolbe's impressively designed Chemical Institute at Leipzig, saw immediately that the City & Guilds' laboratory plans for the Cowper Street playground site were grossly inadequate. Not only would the building be too small to meet the likely student demand, they predicted, it would also be unworthy in comparison with the facilities available in other countries. The Cowper Street school's architect was Edward Clifton (1817–89), a former railway surveyor who had done much work in the City of London with William Tite.[17] Clifton had designed Gresham House and the East India House before erecting the school in 1869 and, on Playfair's recommendation and Wormell's prompting, included laboratories and workshops for mechanics and chemistry teaching. It was Clifton who was initially given the task of designing a cheap two-storey laboratory complex for the City & Guilds, to supplement the school's facilities.

Well over 100 students — bankers, builders, engineers, insurance company clerks, chemists and druggists — attended Ayrton's and Armstrong's classes in the first few months. Armed with this evidence, and with the support of Robins, who was acting in his capacity as the Dyers' Company architect and surveyor, Ayrton and Armstrong were able to persuade the City & Guilds to erect a much larger and better-equipped building in the school's playground on Tabernacle Row (now Leonard Street). The Drapers' Company gave £10 000 for this purpose. As Ayrton later acknowledged, it was because Robins 'strenuously exerted himself to further technical education in Finsbury, that the various electrical, physical and mechanical laboratories now in Leonard Street, Finsbury, came into existence'.[18] Indeed, it was Robins's report to the Guilds on 31 December 1880, in which he over-optimistically argued that a middle-grade technical school for engineering and applied art could be erected in the Cowper Street school grounds for £12 000, that persuaded the Guilds to proceed.[19] Much of his 'clout' undoubtedly stemmed from the fact that he was able to argue that the fittings of a metropolitan technical school ought to be as good as those the Merchant Venturers of Bristol had approved Robins to design and include in their great building at exactly the same time as the London deliberations were in progress.

Although Clifton remained entirely responsible for erecting the building, it is clear that Robins was principally responsible for fitting it out, and that his ideas and designs came from conversations with Ayrton and Armstrong and from personal visits to Mason's College, Birmingham, and Owens College, Manchester, and from the examination of A. W. Hofmann's designs for the chemical laboratories at Berlin (1865) and of the facilities of the Japanese Engineering College.[20] In this way, at a cost of £35 000 (fabric, £21 000; fittings, £15 000[21]), the City & Guilds found themselves committed to building two colleges instead of one.

The foundation stone of England's first technical college was laid at Finsbury by Queen Victoria's youngest son, Leopold, in May 1881. However, the building was not ready for use until February 1883, because legal, labour and cash-flow problems caused delays. Teaching, therefore, continued in the Cowper Street School premises when the first 100 day students began at 'The City & Guilds of London Technical College, Finsbury' in October 1882. Three further significant steps were taken during this interim period. First, at Ayrton's and Armstrong's suggestion, it was decided, in March 1880, to admit day-release students from local factories and (with Wormell's enthusiastic support) from Cowper Street School itself. (One such pupil was William Taylor, FRS, the instrument designer.) Second, in December 1881, Armstrong, Ayrton and Robins went on an extensive tour of the continent to examine laboratory fittings. This tour provided Robins with much of the information he included in *Technical School*; it also gave his views extra weight with the City & Guilds sub-Committee.[22] Expenditure on fittings did lead to some ill-feeling, both between Armstrong, speaking for chemistry, and Ayrton (and Perry), pushing for electrical fittings, and within the sub-Committee of the City & Guilds, who wondered whether their professors were spending for their own research satisfaction rather than for strictly necessary teaching purposes. Although Robins played umpire, the possibility that expenditure on fittings might get out of hand was undoubtedly a major factor in the decision to appoint Philip Magnus as a temporary Director of the new College in 1882 over Ayrton's and Armstrong's heads.[23] Third, in 1882 John Perry was appointed Professor of Mechanical Engineering, with additional responsibility for mathematical instruction. Perry, an unconventional, but brilliant, engineer and teacher, had collaborated with Ayrton in Japan on telegraphic problems and on the design of electrical measuring instruments. Hence, the new building, designed to house two principal disciplines, was forced to do duty for three, with consequent severe pressure on teaching space.[24] Lecture rooms had to be shared, and the College had to make do for much of its existence without waiting or committee rooms, main library or restaurant (the Cowper Street School facilities were used).

Additional evening courses were also added in 1882 when a Department of Applied Art was taken over from the City School of Art, and a Building Trade class was absorbed from the Artizans' Institute in St. Martin's Lane. Applied Art continued to be taught in the Cowper Street School basement until 1891, when it was moved into temporary premises in Leonard Street until the College's extension was opened in 1906. However, the Department was closed in 1912. The Building Trade course was ended in 1899, when it was transferred to the Shoreditch Technical Institute under the auspices of the London County Council's (LCC) Technical Education Board.[25]

The minimum school-leaving age was 12 until 1918, and there was no compulsion to provide 'secondary' education until the Education Act of 1902. Thus, the College's admission of day students (distinct from its evening class work) introduced a new kind of post-elementary school — the technical school or college — into the English education system. The College accepted students from the age of 14, although records suggest that the average age at entry for the day students was 17. Although some of its students went on to the Central Institution at South Kensington for more advanced training to become captains of industry, Finsbury College's more modest and successful intent was to produce foremen, sub-managers and the men and women who planned to take intermediate posts in industrial works. Nevertheless, a number of distinguished chemists and engineers received their first training there: H. A. Crompton, W. J. Pope, G. T. Morgan and J. Read (Professors of Chemistry at Bedford College, Cambridge, Birmingham and St. Andrews, respectively), Sir Alfred Chatterton (Professor of Engineering at Madras and a distinguished writer on Indian affairs), Cecil H. Desch (Professor of Metallurgy at Sheffield and later Superintendent of Metallurgy at the

National Physical Laboratory), William Taylor (founder of the Leicester optical instrument company, Taylor, Taylor & Hobson), H. A. Humphrey (the inventor of the pump), R. W. Paul (a scientific instrument maker and pioneer of cinematography), Julian L. Baker (a distinguished brewery chemist), John H. Coste (chief environmental chemist for the LCC from 1894 until 1936), Sir Henry Royce (the inventor of the famous engine), and the aircraft designers Sir Richard Fairey and Sir Frederick Handley Page. One of the first women students was Hertha Marks, who married Ayrton in 1885. She narrowly missed the distinction of becoming the first woman Fellow of the Royal Society in 1902 for her work on the electric arc, because of supposed legal difficulties in defining 'fellow' to include the female sex.[26]

The four-storeyed plain-brick building contained 32 rooms and occupied 669 m^2 (Figure 9.1). It was fitted with Edisonian electric light powered by a dynamo in the basement, which was driven by a fourteen-horsepower stationary steam engine which also provided motive power to various machines throughout the building by belt drive. Fresh air was drawn from the Cowper Street school playground by fan and passed through a huge heating chamber, from where it was distributed under pressure throughout the building by means of vertical wall shafts and adjustable wall grills. Initially it was intended to ventilate the building by means of a hot-air chimney (as in David Boswell Reid's design for the rebuilt Houses of Parliament in 1838),[27] which was also intended to exhaust foul air from the second-storey chemistry department's fume-cupboards by means of a complementary down-draught shaft. However, as Armstrong — its chief advocate — discovered, the system would not work efficiently, because the main chimney was insufficiently heated, and so, to ventilate his laboratories, a two-foot Blackman's electric fan was placed upon the roof, the ventilation system being taken directly from contemporary mining technology.[28] Although an efficient method, warming the building to some 16 °C, the system suffered the disadvantage of combining ventilation with warming: more air could only be released into a room by raising its temperature. Dust was also a severe problem until the warm-air ducted heating system was replaced by steam radiators in 1908. Other technical colleges and schools built before 1900 learned much from these failures.

The white-bricked basement (Figure 9.2) also functioned as space for teaching mechanics, the design of dynamos (these were driven by a gas engine), plumbing and metalwork. Like the Royal Institution, storage space was also gained by running vaults under the pavement. Vibration of the physics laboratory's stone benches on the ground floor above was also inhibited by piering them to brick wall in the basement's foundation.

The ground floor was entered from the street by a classical portico with a stone frieze bearing the figures of Newton, Wheatstone, Faraday and Liebig. It included three physics laboratories (including one for sensitive experiments), a workshop for testing electrical instruments (Figure 9.3), a mechanics' drawing office, a thirty-seater lecture room and a 'special operations

Figure 9.1 Finsbury Technical College, 1883–1926. The building remains and is used as Shoreditch County Court. From Imperial College Archives.

room' for industrial chemistry (chiefly for the demonstration of sugar refining, brewing and dyeing). This room's exhaust-hoods were fed directly into the main chimney flue. Two private rooms on this floor were given over to Ayrton and Perry, although, in the absence of sufficient administrative space, caused

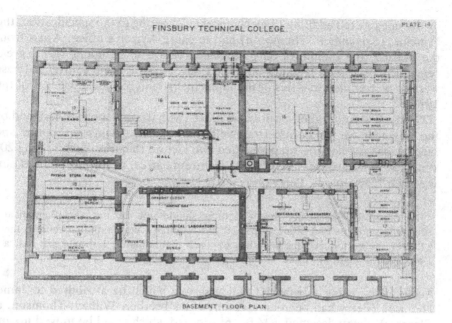

Figure 9.2 Finsbury Technical College, basement plan. From Robins (1887).

Figure 9.3 Finsbury Technical College. A laboratory for first-year electrical engineering students. From *City & Guilds of London Institute Reports and Programmes, 1884–97*, Vol. 1, p. 25.

by the unexpected addition of mechanical engineering to the specification, the Principal (Sylvanus P. Thompson) had to share Ayrton's office. Apart from the portico, the building's only other concession to grandeur was a great central staircase with landings on three sides. A portico and central staircase were to be common features of the municipally built technical colleges of the 1890s.[29]

On the first floor, another drawing office (with drawing boards designed by the engineer Sir Frederick Bramwell) and special rooms for optics and mechanical engineering supplemented two lecture theatres, each seating 200 students. A notable feature of these lecture theatres, as Robins pointed out, was their blackboards. Perry observed:

> It is absolutely necessary that a lecturer should not rub out a mathematical formula until the end of a lesson, and this requires a very long blackboard, the longer the better. As one who has to teach mathematics, I should say that a blackboard ought not to be less in length than 30 feet.[30]

Accordingly, using a balanced pulley system which he attributed to James Thomson (Perry had been an assistant of his brother, William Thomson, at Glasgow), Perry designed a 14 ft × 6½ ft board which could be moved up and down with the touch of a finger. The mechanism was left exposed so that auditors were provided with a visual demonstration of mechanical principles. Such blackboards are, of course, a commonplace of twentieth-century teaching institutions.

No student common-room was provided initially; instead, space under the tiered lecture rooms was used for this purpose. The second floor also housed chemistry and physics preparation rooms adjacent to the lecture theatres, for the three professors still took the art of lecture demonstration very seriously.

The second and final floor was devoted entirely to chemistry (Figure 9.4). Two asphalt-floored laboratories (they were separated only by iron columns supporting the roof) catered for 96 junior and senior students, but, because the benches had storage space for two students, the effective capacity was for 192 students working in two shifts. Armstrong noted the deliberate difference from continental practice, where advanced students doing quantitative analysis were physically separated from junior students engaged in qualitative analysis. The Finsbury practice made it easier to supervise students — an important factor, because staffing levels were much lower in Britain than on the continent.[31] Armstrong and Robins solved the perennial problem of ventilation in chemistry laboratories by running hoods along the lengths of the junior students' benches.[32] Senior chemists had access to wall fume-cupboards which were improvements by Armstrong of Hofmann's original gas draught design.[33] As mentioned, the system only worked efficiently after electric-fan suction replaced the down-draught chimney principle of the original design (Figure 9.5). Natural light was enhanced by roof glass. An unusual feature of the junior laboratory, the benches of which ran parallel to

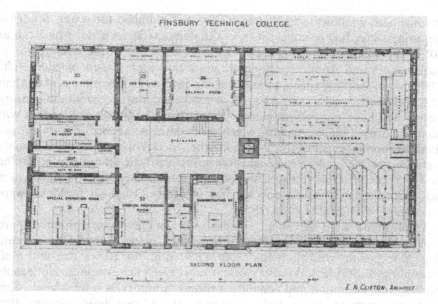

Figure 9.4 Finsbury Technical College. Second-floor plan showing Chemistry laboratories. From Robins (1887).

Figure 9.5 Finsbury Technical College. Chief Chemistry Laboratory on the north side of the top floor. From City & Guilds of London Institute Prospectus form 1912–13.

the northern windows, was that they were not plumbed for water. Students requiring water had to collect it from the demonstration table, which ran at 90° to the windows and was under the stern discipline of the demonstrator. Armstrong's and Robins's ergonomic ideas that each student needed 30 ft^2 (2.78 m^2) of working space, so that benches had to be a minimum of 4 ft 6 in (1.37 m) apart, to allow safe passage, were handed down to twentieth-century design and have not been replaced.[34] Armstrong's third-floor kingdom was completed by a balance room (which doubled as a chemical library); a gas analysis room; a small lecture room; another special operations room, mainly used for organic preparations; and Armstrong's private laboratory, where he had 4–5 places for his research students. Given these excellent facilities, it is not surprising that Armstrong and his successor, Raphael Meldola, were extremely productive in research.[35]

Unplastered walls served to emphasize a workshop atmosphere of hard work, for students were expected to be in their places from 9.30 a.m. until 5.30 p.m., with a lunch break from 12.30 p.m. to 1.15 p.m., and staff (including the professors) also worked most weekday evenings. Noticeably absent were reception and committee rooms, a library, a refreshment room and a staff common-room. Indeed, by oversight, a ladies' toilet was forgotten and had to be placed on the roof above the second-floor Gents!

The City & Guilds' resources were considerably stretched by the cost of building and fittings of Finsbury College (£35 000). This was especially so when the Drapers' Company, angered by the decision to develop the Central Institution outside the city, decided not to put further money into the co-operative venture. In 1889 the Charity Commissioners suggested that Finsbury College might amalgamate with the Northampton Polytechnic (now the City University), which they proposed to build less than a mile away; however, the threatened loss of the Livery Companies' capital investment prevented this being considered seriously.[36] But by 1920 the Northampton Polytechnic Institute was a thriving rate-supported institution with a large up-to-date engineering school modelled on Finsbury's course by its Principal, Robert Mullineux Walmsley, a former electrical demonstrator at Finsbury College.[37] Despite a grant-in-aid from the LCC in 1921, after much agitation and protest Finsbury College closed its doors in 1926.[38] Its staff and students had been depleted during World War I, and its expenditure was greatly out of step with its income from fees. Moreover, it was now surrounded by better-endowed competitors whose efficiency had been modelled upon it, while the 1918 Education Act had imposed a duty upon all local authorities to provide comprehensive education (including technical) in their areas. The main building still survives as a County Court for Shoreditch.

What had the 7000-odd engineers and technicians it trained gained from their education in its laboratories?[39] Its teachers, old students and contemporary observers all spoke of a 'Finsbury method' or 'plan'. Five features of the 'method' can be characterized.

First, the teaching in all three departmental areas of chemical, electrical and mechanical engineering was analytic (or, in Armstrong's terminology, 'heuristic') rather than synthetic or deductive.[40] A fundamental principle was that students should be taught to think for themselves.

Second, practice in the laboratory and workshop was more important than lectures. Both Ayrton and Armstrong developed printed instructions to accompany experiments.[41] These raised questions the students had to answer through their own manipulations. Moreover, experiments were usually quantitative in character, thus raising the issue of functional relationships and the need to develop mathematical skills of analysis. Through Armstrong, 'learning by doing' was developed into an educational movement for reforming and revitalizing school science teaching.[42]

Third, the development of a fresh 'practical' mathematics syllabus, which recognized that the mathematical needs of scientists and engineers were different from those of potential mathematicians, became an important feature of Finsbury's programme. Here Perry developed *Mathematics for Engineers*, a best-selling text which emphasized decimals with approximations and the use of logarithmic tables and the slide rule. In algebra the use of formulae without derivation was emphasized and functions were studied for the first time in Great Britain by graphical methods using squared paper.[43] Practical methods in mensuration and geometry were encouraged, as well as numerical trigonometry. All these elements were variously mixed together, while a close correlation was kept with the students' science lectures and laboratory work. This sense of a broader-based mathematics curriculum, together with the idea of 'mathematics through the engineering curriculum', was to become extremely influential in discussions of the British school mathematics and science curricula during the Edwardian period, in which John Perry and Henry Armstrong played leading roles.

Fourth, all students, whatever branch of engineering they wished to enter, took a common first-year course which included chemistry, mathematics, mechanical drawing, electrical and mechanical engineering, and French or German. As Armstrong commented on his own discipline, 'the introduction of mechanical drawing and the elements of engineering practice into the chemical course was a departure of great consequence, which has given Finsbury students a special value in works'.[44]

Finally, although students were selected by an entrance examination consisting of papers in English language and mathematics (or, alternatively, were required to have passed London University Matriculation standard), this condition was mainly imposed to stimulate local schools to the high standard already achieved by their neighbour, the Cowper Street Middle Class Boys' School. For the Finsbury philosophy was firmly set against any form of outside examination. Students registered for a two-year course whose end product was a diploma, and the college opposed competition with the University of London.[45] The chemistry course was extended to three years in 1899 because of the demand for fully trained chemists by industry. The two other engineering departments followed suit in 1906, when an extension, designed by Robins as early as 1886,[46] was erected on the site of a timber yard adjacent to the College. This extension (now a factory) provided a mechanical engineering office and lecture theatre, a library, an enlarged mechanical workshop and proper studios for the Department of Applied Art.[47]

The policy of ignoring outside examinations had the great advantage that syllabuses could be very quickly altered as technology improved or changed. For example, when during the Edwardian period the demand for electricians diminished, Finsbury was able to produce more engineers capable of dealing with the new technology of the internal combustion engine. Certainly, it was not something that could have been done quickly if, say, the college had been tied to a London university examinations system.

A mark of the originality of the Finsbury plan is that so many of its features now seem commonplace. Finsbury succeeded in its intended task of training technicians for Britain's 'second industrial revolution' — a revolution based not upon steampower but on chemicals and electricity and the petrol engine. And through its factory-like architecture, its distinctive curriculum and philosophy of the laboratory and workshop, it exerted a profound influence on the development of British technical education. In the words of one of its alumni, 'Finsbury blazed a trail which many have followed'.[48] That the college closed in 1926 was not a mark of failure but a sign that its work as a trail-blazer was done.

Notes and References

1. Robins (1887), dedicated to T. H. Huxley. There is some overlap with the earlier volume of reprinted papers and addresses (Robins, 1885a), which was dedicated to the engineer Sir Frederick J. Bramwell, the first Chairman of the City & Guilds.
2. Smith (1976); Cresswell (1975); and, generally, Lowe and Knight (1982). Webb's academic buildings included the Royal College of Science (Imperial College, 1900–6), the Victoria and Albert Museum (1891), Christ's Hospital, Horsham (school, 1894–1904) and the University of Birmingham at Edgbaston (1906–9).
3. Clay (1902); Russell (1903); Munby (1929); Jenkins (1979), pp. 252–89, 'Laboratory provision and design'.
4. I have assembled details of Robins's career from a notice in *Building News* (1890), **58**, 793 (portrait); an obituary in *J. Roy. Inst. Br. Architects* (1917–18), **25**, 205–6; and *Robins's and Robson's. A Bibliographical Summary of Their Work*, duplicated handlist, RIBA, 1958–9.
5. Class rooms which open directly onto an assembly hall. The system seems to have been hit upon independently by several architects in the 1880s, who were influenced by the separate classroom system pioneered by the Prussian government. See Robins (1887), pp. 202 *et seq.*, and Seabourne and Lowe (1977), pp. 25–7, 51.
6. Republished as Godwin (1854, 1859). Robins's reminiscences are made in his (1881). See also *Report of Local Board of Health for the District of Regent Square Church, St. Pancras, a Practical View of the Sanitary Question*, 1854 — copy RIBA.
7. Morley (1971), Chap. 3.
8. Wohl (1983), p. 145.

9. Livery Companies' Committee, *Report on Technical Education*, 1878; Foden (1970); Lang (1978).

10. (Pingree, n.d., 1974). Note especially Robins to Armstrong, 25 February 1887, where Robins asks Armstrong to arrange his election to the Royal Society for his services to architectural science.

11. Robins (1887), p. 2.

12. From 1893 to 1907 the Central Institution was known as the Central Technical College — whence the title of the College's informative magazine, *The Central*; and from 1907, when it became part of Imperial College, it was officially designated the City & Guilds College.

13. Harte (1986), p. 187.

14. Supported by Lyon Playfair, Wormell had experimented with evening classes on technical subjects during 1875, but found them unprofitable. See undated letter [1880?], Wormell to City & Guilds Committee, City & Guilds archives, Box E11. On the important Cowper Street School (founded 1866), now the City of London Boys' Foundation Grammar School (and still in its original premises), see Bryant (1986), pp. 238–47.

15. Ayrton (1892).

16. Brock (1981).

17. Obituary in *Builder* (1889), **56**, 34.

18. Ayrton (1892), p. 8. For Ayrton and Armstrong's desiderata, see *Central* (1938), pp. 17–24.

19. City & Guilds of London Institute, *Report to the Governors for the Year Ending March 10th 1880* (copy City & Guilds archives). Robins's recommendations, pp. 42–5. Also *Report . . . Nov. 8th 1880*. Note that Robins was also deeply involved in the building of the Merchant Venturers' College, Bristol, at this time.

20. Robins (1887), pp. 75–6, 139–40, 152–3, 173–5. Cf. Armstrong's letter, 9 June 1882, complaining of Clifton's slovenly brickpointing and carpentry (City & Guilds archives, AFB/1029). Clifton had visited Mason's College, Birmingham (extensions designed by J. A. Cossins, 1874–5), before making his design, but Armstrong was appalled to find the chemistry laboratory only measured 30 ft × 40 ft. Robins's improved design was 50 ft × 42 ft.

21. This was half the cost of Mason's College (£67 000) and only a third of the cost of the Central Institution (£100 000). Figures from Robins (1887), Appendix, which also gives building sizes and maintenance costs for these and many other contemporary British science laboratories. In 1885 the Drapers' Company awarded Finsbury College a further £10 000 for equipment. See Thomson [1901], p. 9.

22. Robins to Armstrong, 4 February 1882 and 31 July 1883 (I.C. archives); Robins (1887), pp. 17–20, 43–70.

23. *Ibid.*, p. 8, and Robins to Armstrong, 13 November 1882 and 15 November 1883 (I.C. archives).

24. Between 1891 and 1906 a small warehouse was rented in Leonard Street to house a heat laboratory and the mechanical engineers' drawing office.

25. City & Guilds, *Report . . . 14th March 1883*; Millis (1925).

26. Sharp (1926). My account of Finsbury College draws upon: Magnus (1883); S. P. Thompson [1901]); *Royal Commission on University Education in London. First Report*, appendix, 'Finsbury College'. *Pml. papers* Cd. 5166 (1910); Streatfield (1912); Walker (1933, 1950); Armstrong, H. E. (1934); Armstrong, E.F. (1938)); Allam (1952); Lang (1978); Eyre (1958), pp. 67–84, 290–3; Jordan (1985). Walker (1878–1954) was one of the first English consulting engineers to specialize in concrete. For Paul, and other Finsbury graduates, who joined the Cambridge Scientific Instrument Company, see Cattermole and Wolfe (1987), pp. 98–105, 142–3, 215–16.

27. Port (1976), Chap. 11. Continental trips with Armstrong had made Robins aware that 'no professionally-recognised system for heating and ventilating public buildings existed in this country, and I resolved to embrace the first opportunity which occurred to introduce . . . the principles in vogue in Belgium and Germany'. Robins (1885b). Cf. Glasgow's new physical laboratory (1870), where fresh air was drawn down a shaft by fans worked by a basement steam engine: *Nature, Lond.* (1872), **6**, 29–32.

28. Robins (1887), pp. 174–5, quoting Armstrong; Robins (1885a), p. vi, and (1882–3), pp. 99–100. The main reason for experimenting with down-draught chimneys was because of the problem of chemicals condensing when drawn directly upwards. See Robins (1887), p. 89.

29. Architecturally, Finsbury seems to have particularly influenced the design of Heriot Watt College, Edinburgh (1885–9), Leicester Technical College (1896) and Sydney Technical College (1891), on which see Magnus (1910), pp. 84–127, especially p. 90.

30. Robins (1887), p. 152.

31. Armstrong's comment on Robins (1882–3) was not included in the reprint of this paper in Robins (1887), pp. 71–114.

32. A similar system was used in the 1840s in the basement laboratory of the Pharmaceutical Society.

33. Robins (1887), Plates 47 (Hofmann), 50 (Armstrong); and pp. 123–4, 152.

34. Jenkins (1979), p. 259. Adopted in Board of Education *Regulations* (1907).

35. Eyre (1958); Marchant (1916). Meldola's successor, G. T. Morgan (a former student at Finsbury), held the chair from 1915 to 1919 before going to the University of Birmingham, where he did his main work on chelation. The final chairholder, A. J. Hale, was undistinguished in research, but compiled chemistry texts for engineers and Hale (1921–5).

36. *Report . . . 1889*, p. 8.

37. Teague (1980).

38. The City & Guilds announced the College's closure in July 1921 and recruited no first-year students for 1921–2. After protest, and the LCC grant, new students were recruited for 1922–3.

39. There were 7180 day students between 1884 and 1926 (3069 electrical engineers and 1819 chemical engineers). See *Report . . . 1926*.

40. The Department of Applied Art does not interfere with this point. For a good treatment of Ayrton's 'analytical' method, see Jordan (1985).

41. Brock (1980). See also *Memorandum of Proceedings at a Drawing Room Meeting for the Promotion of Technical Education Held at the House of Mr. E. C. Robins . . . 5th March 1887* (privately printed), p. 14. Copy Imperial College Archives. I am grateful to Dr Sophie Forgan for this reference.

42. Brock (1973).

43. Brock and Price (1980); Price (1983).

44. Armstrong (1916), p. 87.

45. Thompson [1901], pp. 2–3.

46. City & Guilds archives, Box E11.

Section 3

THE LARGE PHYSICS

LABORATORIES

Abbreviations

AEC	Atomic Energy Commission
AG	alternating gradient
AGS	Alternating Gradient Synchrotron
AR	Accelerator Research
AUI	Associated Universities Inc
BeV	Billion electronvolts
BNL	Brookhaven National Laboratory
CERN	Centre Europeenne pour la Recherche Nucleaire
FFAC	Fixed Field Alternating Gradient Synchrotron
GeV	Gigaelectronvolts
HEP	High-energy Physics
ISR	intersecting storage rings
JCAE	Joint Committee for Atomic Energy
LRL	Lawrence Radiation Laboratory
MED	Manhattan Engineer District
MeV	Megaelectronvolts
MSF	Million Swiss Francs
MURA	Midwest Universities Research Association
NAACP	National Association for the Advancement of Colored People
NAL	National Accelerator Laboratory
NAS	National Academy of Sciences
NSF	National Science Foundation
PS	Proton Synchrotron
SC	Synchro-cyclotron
SPC	Scientific Policy Committee
TNL	Truly National Laboratory
URA	Universities Research Association
WAG	Western Accelerator Group

10

Pragmatism in Particle Physics: Scientific and Military Interests in the Post-war United States

Andrew Pickering

Developments in elementary-particle theory in the post-World War II era point to a novel instance of the Forman thesis, with an ironic twist. In his well-known paper on 'Weimar culture, causality, and quantum theory, 1918–1927', Forman argued that the admission of acausality and indeterminism into theoretical physics — a key step in the formulation of modern quantum mechanics — could be related to Germany's defeat in World War I. The intervening variable was the hostile, anti-scientific and anti-deterministic intellectual milieu which grew up in the Weimar Republic, and to which German theorists accommodated themselves. The distinctive, and disturbing, aspect of Forman's argument was that in the formulation of quantum mechanics gross and profane worldly events — namely, a military defeat and its social repercussions — somehow inserted themselves into the sacred texts of theoretical physics, albeit in highly encoded form.[1]

Undeterred by the mixed reception which Forman's argument has received, a novel instance of the same theme has recently been taking shape in the work of several historians of modern physics, most notably that of Silvan Schweber.[2] Here the emerging argument is that not defeat but military victory left its mark on the sacred texts produced in the late 1940s and through the 1950s. The victory was that of the United States in World War II and, once again, the intervening variable was the resulting social milieu. Unlike the situation in Weimar Germany, this milieu was, of course, highly favourable to science, but it would seem that the heart of theoretical physics is as easily swayed by adoration as by hostility.

My aim here is to spell out the structure of this latter-day version of the Forman thesis. I begin by discussing theory development in US high-energy physics (HEP) in the 1950s. I argue that this was characterized by a distinctive pragmatism, and I further argue that this form of pragmatism was not dictated by the subject matter of the field: it stands in need of explanation. For understanding, I look to the social milieu within which the rapid growth of HEP was sustained. I suggest that HEP was nurtured by its symbiosis with powerful political and military interests, and I sketch out some of the processes whereby this symbiosis structured theory production within pure science.[3] Although, as stated, I am primarily interested here in theoretical rather than experimental physics, key sites of this symbiosis were, of course, the big-science laboratories for HEP experiment that grew up after the war.

What was distinctive about US HEP theory in the post-war period? Let me discuss the early development of 'dispersion relations' as a typical example.[4] For present purposes, the history of dispersion relations begins in 1955, with the publication of Marvin Goldberger's 'Causality conditions and dispersion relations'.[5] In that paper, Goldberger argued that certain general relations should exist between the quantum mechanical amplitudes which describe the scattering and interaction of elementary particles. In particular, the real parts of these amplitudes should be given by weighted integrals over the imaginary parts. These integral relations were the dispersion relations. From Goldberger's work there followed, in his own words, 'an explosion of activity': dispersion relations dominated strong-interaction theory in HEP for the remainder of the 1950s and through the 1960s.[6] Here I want to explore what lay behind the popularity of dispersion theory and characterized its content.

Two features of Goldberger's 1955 publication should be noted. First, although grounded in the traditional approach of quantum field theory, Goldberger's derivation of dispersion relations was at best heuristic rather than rigorous. In his own words again: 'Of course, my derivation was not really correct; [but] since the result was correct, I was sure that someone would prove it eventually.'[7] This cavalier attitude towards formal theory was one distinguishing mark of the dispersion-relation programme. Proofs of Goldberger's original conjecture did eventually appear, but many heavily used extensions to new domains of application were never certified by field theorists.

Reference to use and application brings us to the second distinguishing mark of the dispersion-relation programme: its relation to experiment. Despite their lack of rigour, dispersion relations did mark a significant point of contact between quantum field theory and high-energy experiments. Dispersion relations could be formulated in terms of measurable quantities in elementary-particle interactions, and had an immediate engagement with HEP data which had been lacking in earlier attempts to apply field theory to the analysis of 'strong-interaction' processes.[8] Dispersion relations, then, constituted a set of phenomenological tools available for the analysis of the masses of data on strong-interaction physics which emerged during the 1950s from the post-war crop of high-energy particle accelerators. One could, for example, attempt to use experimental data to test the validity of dispersion relations (and, hence, the validity of the putatively underlying field theory); alternatively, one could assume the validity of the relations themselves and use them to extract additional information from data: to choose between alternative phase-shift analyses, or to extract coupling constants.[9]

These two characteristics — a split from fundamental theory, coupled with phenomenological utility — define what I refer to as the pragmatism of post-war HEP theory. I want now to indicate why this pragmatic orientation stands as problematic, before outlining the elements of an explanation.[10]

At least three observations point to the conclusion that pragmatism was not a *necessary* feature of post-war theory. First, as Schweber has shown in a comparison of the 11th and 12th Solvay Conferences, held in 1948 and 1961, respectively, the pragmatic approach marked a clear break from historical precedent. The 1948 conference was dominated by the pre-war European élite, and focused on formal theoretical topics divorced from experiment; the 1961 conference was instead dominated by the young American élite, and focused on the applications of approaches such as dispersion relations to experimental data. Second, while the pragmatic approach was dominant in theory, other styles of theorizing remained evident. Thus, for example, while most theorists followed Goldberger's example and cared little whether theoretical justification would follow phenomenological application, a small but distinct and primarily European group of theorists pursued the problem of rigorously establishing the existence of dispersion relations in quantum field theory. This was the tradition of work that became known as axiomatic field theory.

Third, as is already implicit in the above, the pragmatic approach flourished above all in the United States. While the phenomenological utility of dispersion relations was being energetically exploited by US theorists, theorists on the other side of the Atlantic remained preoccupied with more formal and fundamental questions in quantum field theory. It is clear, therefore, that the pragmatism typified by the dispersion-relation programme was not a necessary or inevitable feature of post-war theory: there were alternative ways of conceptualizing the subject matter of HEP historically available and present in contemporary practice; the US theoretical community chose by and large to ignore them.[11]

The pragmatism of post-war US theory thus constitutes a historical problem. I want now to argue that it is to be explained in terms of the interweaving of scientific, political and military interests within post-war American society. Two points must be emphasized in advance. First, much of the documentation and analysis to be summarized is not the product of my own research: it comes from the work of the historians cited. Second, many details of the historical picture remained to be filled in. My justification for what follows is a belief that the argument is important, and that it stands at present in need of clarification. I shall therefore present my perception of what is at issue as starkly as I can, on the understanding that more work is needed on almost every point.

World War II, 'the physicists' war', marked a discontinuity in the history of both physics and the military, at least in the United States, in which each surrendered a degree of its traditional autonomy in exchange for benefits the other could provide. During the war the armed forces reformed themselves to put physics at their centre, reluctantly at first but later, in the light of operational successes, ardently. The new technologies of, for example, radar and atomic weaponry increasingly dominated military thought and action.[12] One consequence of this reformation of the services was an increasing dependence upon what I shall call advanced physical *technique*, referring both to the military *hardware* that was the product of basic research and to a new form of scientific life embodying *expertise* in the development and use of such hardware.[13] This new form of life was itself born in the military laboratories of World War II: laboratories where large, hierarchically organized teams of physicists and engineers came together in capital-intensive projects organized around technologies of use to the military. The establishment and maintenance of such laboratories — the MIT Radiation Laboratory, Lawrence's Radiation Laboratory at the University of California, Berkeley, and Los Alamos being prime examples — was expensive, and the new form of scientific life that was born in those laboratories could only survive at the unprecedented funding levels that the services were willing to provide. And thus, as the military became dependent upon physics in World War II, so the new form of technique-oriented scientific life born dependent upon the military.[14]

In the years following World War II the mutual embrace of physics and the military initially relaxed, but was only reinforced in the era of the Cold War, especially after the United States became involved in the Korean War in 1950. Here US politicians and military men perceived themselves to be locked into a technoscientific arms race with the Soviet Union, which they could only win through a continuing relationship with advanced physical technique: the precondition for success in the arms race was seen to be the propagation forward in time, the reproduction and multiplication, of the scientific form of life that had proved so useful in war — technique-oriented physical research, centrally organized around particular material apparatuses and phenomena of potential military significance.

Not surprisingly, in the 1940s and 1950s the most significant techniques to the military mind were the descendants of the World War II success stories — the techniques of electronics and nuclear and particle physics. And these were the fields most heavily sponsored by 'defence'-related agencies. In the case of HEP, funding and material were forthcoming from the Manhattan Project and, from 1946 onwards, from the Atomic Energy Commission for the construction of major accelerators at Lawrence's Radiation Laboratory in Berkeley and later at the Brookhaven National Laboratory on Long Island, as well as for smaller projects elsewhere. Equally unsurprisingly, the big-science descendants of the wartime projects were eager to accept this largesse: fields such as HEP never escaped from dependence upon massive federal funding.[15] But it is important in what follows to bear in mind that this continuing embrace of HEP and the military in the post-war era rested upon

an intersection rather than ar / necessary equation of interests. The military
was interested in the continuing health of HEP as a source of technique, for
present and future military purpses, rather than as a source of knowledge
about the innermost constituents of matter. High-energy physicists, in
contrast, although they frequently did share this military interest, could
legitimately claim that the latter form of pure scientific knowledge, having no
conceivable practical utility, was the only true product of their work. The
HEP laboratories at Berkeley, Brookhaven and elsewhere thus marked a
cut-out as well as a point of intersection between physics and the military. A
bargain, first partly explicit but late: largely implicit, had been struck, in
which both generals and physicists could get what they wanted — the
multiplication of technique and knowledge of the constituents of nature,
respectively — while apparently standing quite apart from one another.[16]

With this background in mind, we can start to think about the pragmatic
post-war approach to physical theory. I want now to unravel some of the ways
in which the wartime and post-war US military investment in a particular
form of scientific life contributed to the development of theories such as
dispersion relations.

The first point to note is that pragmatic theorizing was an entirely
appropriate mode of thought to the wartime military laboratories in which the
marriage between physics and the military was consummated. The aim of
these laboratories was to produce hardware that worked — bombs and radar
sets — not fundamental knowledge about physical processes. Acceptance of a
divorce from fundamental theory as the price to be paid for useful knowledge
— knowledge that could guide and structure experimental research and that
could give sense to its findings — was as characteristic of the Manhattan
Project as it was of dispersion relations a decade or so later. This isomorphism
was not a coincidence. The post-war élite of US HEP theory was largely
drawn from physicists who had first practised their trade in the overridingly
pragmatic context of wartime military research. This élite — Goldberger
among them — took their habits and routines of theory development with
them from war to peace, from the design of atomic bombs to the interpretation of pion–nucleon scattering.[17] Furthermore, as an élite, the veterans of
World War II were in a position to enforce their own work style. They trained
the next generation of theorists, and they held institutional power — playing a
dominant role in determining, for example, what should be published, who
should be hired, what should be on the agenda for the next international
conference and thus what was on the agenda for elementary-particle theory as
a whole. In short, the élite was able to *reproduce* the theoretical culture which
had evolved at the wartime laboratories: Chicago, Berkeley, MIT, Los
Alamos.

In considerations such as these one can see the outline of an explanation for
the pragmatism of US theory in the late 1940s and 1950s. But at this level the
explanation is a superficial one. Questions remain: Why did the élite remain
faithful to its wartime orientation, rather than returning to more European

pre-war working practices? More profoundly, how and why was this particular élite and its distinctive pragmatism constituted and sustained? To answer such questions one must look to the circuits, direct and indirect, that continued to bind HEP theorists to the military.

At the most direct level, the peacetime loyalty of élite HEP theorists to wartime practices was confirmed and rejuvenated by a continuing military involvement. Although the knowledge of elementary particles generated by these theorists constituted a negligible input to 'defence' in the 1950s and 1960s, their expertise was repeatedly drawn upon, and their place in the 'defence' establishment was increasingly routinized in the Cold War era. Schweber has already discussed the involvement of HEP theorists in military projects at Los Alamos, Berkeley, Princeton and elsewhere during the 1950s, and also the institutionalization of such arrangements in the formation of the Institute for Defence Analysis in 1956. Here I want just to emphasize that the continuing experience of practice within a military context was evidently a source of reinforcement for the pragmatism of the existing élite, and an important training ground for new entrants to it.[18]

But what of the theoretical élite itself? How was its stability guaranteed? Clearly, involvement with prestigious defence projects must have played some role, but the question remains of how this particular élite could have sustained its distinctively pragmatic orientation *within* the HEP community. To give sense to this question and to answer it requires a closer inspection of the post-war development of US HEP. We need to bear in mind that HEP theorists were an integral part of a larger community, and to look at indirect circuits leading to the military through the high-energy accelerator laboratories.

As I have already noted, the post-war military interest in HEP was in technique — in, to put it crudely, gadgets of potential future use to the military, and in the maintenance and expansion of a pool of expertise in their use. This interest lay behind the heavy financial sponsorship of unprecedentedly expensive HEP accelerator laboratories in the post-war United States, at Berkeley, Brookhaven and elsewhere. These laboratories were the site of the technique that attracted the military, and their growth, in turn, further reinforced the pragmatism of post-war US HEP theory, along axes that I will now sketch out.

As suggested above, there is no reason to assume that the sponsors of the HEP laboratories had any direct interest in the knowledge of elementary particles to which they contributed. But equally there is no reason to assume that the physicists who worked in them did not. And here there was a problem. The particle accelerators of the post-war era generated immense quantities of data. But data are meaningless without theoretical interpretation, and the interpretative tools inherited and refined by post-war particle theorists — those of perturbative quantum field theory — proved to be blunt instruments for the dissection of weak- and strong-interaction processes, precisely those processes of interest at the big machines. A framework for the analysis of weak-interaction processes was readily achieved which was adequate in the phenomenological sense, if not from the standpoint of fundamental theory, but even this limited achievement eluded theorists of the strong interaction until the advent of dispersion relations.[19]

It thus begins to become clear why the pragmatism of the theoretical HEP élite — typified in the dispersion-relation programme — was welcomed and sustained by the HEP community at large, including the experimenters: theoretical approaches such as dispersion relations elevated the status of high-energy experiment beyond that of mere fact collection. Stated like this, this aspect of the pragmatism of post-war theory is in danger of appearing trivial: of course, data require interpretation. But recall that alternative ways of theorizing were available; theorists were not forced to set up the phenomenological industries that they did. They could, for example, have claimed that their first priority had to be to set dispersion relations on a sound field-theoretic footing before it would make any sense to indulge in phenomenology. To understand why, by and large, such claims were not made, one has to look at the various social contexts in which US HEP theorists were embedded.

First, and most evidently, particle theorists practised their trade in a context partly defined by their experimental colleagues. All of the accelerator laboratories supported theory groups, and success within these groups was defined to a significant extent by the utility of theory to experiment. More generally, experimenters were the peers of theorists within the HEP élite; experimenters had, that is, at least an equal say in determining what constituted useful theorizing.[20] And one can see not one but several reasons why HEP experimenters were inclined to welcome theoretical approaches such as dispersion relations. As noted, theory, however pragmatic, invested experimental data with meaning. Just as important, approaches such as dispersion theory could be used to give structure to extended programmes of experiment.

Here we touch on an aspect of post-war Big Science which has been discussed by Peter Galison. Experimental HEP took over from wartime research a work style quite alien to pre-war physics. The construction and utilization of accelerators and detectors involved long-term collaborative efforts and commitments. Experiments could no longer be performed quickly and cheaply by a single physicist. Within the HEP community, individual experimental proposals had to be argued for and justified, increasingly through institutionalized channels. In this situation opportunities for individual idiosyncracy and initiative were much reduced from the level of the pre-war lone researcher, and the importance of shared theory was correspondingly enhanced as the basis for the evaluation of experimental proposals.[21]

Thus, the dynamics of life within the big laboratories was itself sufficient to foster pragmatism in HEP theory: in structuring long-term collaborative research programmes, in peace as in war, detailed phenomenology was just what was needed. Furthermore, it is important to remember the wider social context in which HEP grew up. The American public was ultimately paying the bills for all of the new accelerators, and some public justification for the expenditure was called for. Typically, then as now, the justification was in terms of the knowledge of the natural world produced — theory was centrally involved. And if fundamental theory had nothing to say on the details of experiment, then phenomenology would have to do.[22]

Thus, the pragmatism of US theorists was sustained by military interests indirectly as well as directly. The growth of the large laboratories was sustained by an overall military interest in the extension and multiplication of technique; the experimental context within those laboratories and the wider public context sustained a demand for useful, pragmatic theorizing.

One last factor pointing post-war HEP theory towards pragmatism needs to be mentioned: the overproduction of manpower. Part of the object of the exercise, as far as politicians and military men were concerned, was to expand the supply of trained personnel with appropriate skills, not that those personnel should remain forever within the HEP community. And, indeed, during the 1950s more high-energy physicists were trained than could find permanent employment within the field. This overproduction had two significant correlates. First, it generated within the community an intensely competitive environment, in which the continual display of original research was a necessity if one were to survive. And here the constant outpouring of experimental data clearly favoured the pragmatic theorists skilled in phenomenological analysis at the expense of, say, the axiomatic field theorist, whose work derived no stimulus at all from empirical input.[23]

The second correlate of the overproduction of manpower is even less subtle. Many trainees in particle physics did not pursue careers in academic research. Instead, they found employment either in military research or in the post-war technological infrastructure which had grown up in association with it. And the skills in demand here were the pragmatic skills of the dispersion theorist, not the formal skills of the axiomatic field theorist. As the élite and many apprentices were surely aware, useful training for those who would one day leave the HEP community — training that would fit them for a career in 'the real world' — was training tailored to the needs of the military–industrial complex. Pragmatism in theoretical physics was, therefore, the best guarantee of employment prospects inside or outside the academic life.

The argument is, then, that in the development of post-war US HEP theory one finds a novel instance of the Forman thesis. Profane military and political interests made their mark upon pure scientific knowledge, sustaining the distinctive pragmatism of post-war theory. I have traced out some of the circuits that bound post-war theory to such interests: direct circuits leading through the wartime laboratories and élite post-war consulting for the military; and indirect circuits leading through the accelerator laboratories. Concerning the laboratories, I have suggested that, on the one hand, they were sustained by a military interest in the reproduction and expansion of the material and social components of advanced physical technique; and that, on the other hand, these laboratories created a context that in various ways further reinforced the pragmatism of post-war theory. Finally, I drew attention to the manpower flows into the military–industrial complex inherent in the military interest in technique, and I noted that these flows, too, reinforced a pragmatic approach to theorizing.[24]

Notes and References

1. Forman (1971).
2. Schweber (forthcoming). See also Cini (1980).
3. My concern here extends only to the end of the 1950s, but it can be argued that the pragmatic style of theorizing institutionalized during this period has continued to dominate HEP to the present. Compare the analysis of the subsequent period developed in Pickering (1984), especially Chap. 12, and in Pickering and Trower (1985) with that of the 1950s offered here and in Pickering (forthcoming).
4. For a fuller account of the history of dispersion relations and references to the secondary literature, see Pickering (forthcoming).
5. Goldberger (1955b). This was the third in a sequence of programmatic explorations in field theory, beginning with Gell-Mann et al. (1954) and Goldberger (1955a).
6. Goldberger (1961), p. 181. See also Schweber (forthcoming), pp. 25ff.
7. Goldberger (1970), p. 688.
8. More direct attempts to apply quantum field theory encountered the problem that traditional perturbative methods were vitiated by the magnitude of the strong-interaction coupling constant; see below.
9. For a fuller discussion of the uses to which dispersion relations were put, see Pickering (forthcoming).
10. For further characterization and documentation of the pragmatism of this period, see Cini (1980), Schweber (forthcoming) and Pickering (forthcoming). There is, I think, a case for seeing all scientific practice as pragmatic, in the sense of being oriented towards particular, situated technical goals, rather than towards transcendental goals such as truth or beauty. I am using 'pragmatic' in a more restricted sense here, as defined in the text.
11. Schweber (forthcoming), pp. 2–4, 22–4; Pickering (forthcoming), pp. 20–1. On the history of axiomatic quantum field theory, see Wightman (forthcoming). Although the topic needs further investigation, it is conceivable that the relative lack of enthusiasm for pragmatic theorizing in Europe correlates with the existence of significant resistance in the European physics community to the new big-science accelerator-based form of HEP life (see below), and that that correlation is itself related to a relative lack of intimacy between the European scientific and military establishments: for hints in this direction, see Hermann et al. (1987), Vol. 1, passim.
12. See Kevles (1978), Chaps. 20 and 21.
13. This sense of technique I take from Ellul (1964); the idea that the post-war military interest revolved around advanced physical technique I take from Forman (1987), especially pp. 220–4.
14. See Kevles (1978). The form of life of the wartime laboratories was not entirely unprecedented: E. O. Lawrence's Radiation Laboratory at Berkeley is one evident pre-war model. What was new in World War II was the exposure of almost the entire US physics community to the Rad Lab work-style on a vastly

expanded scale. For a history of the Rad Lab from its pre-war beginnings, see Heilbron *et al.* (1981).

15. Kevles (1978), Chap. 23. On the post-war history of Lawrence's laboratory, see Heilbron *et al.* (1981) and Seidel (1983, 1986, forthcoming). On the history of Brookhaven National Laboratory, see Needell (1983). For an exceptionally thorough overview of the post-war military dominance of physical research in electronics, see Forman (1987). Leslie (1987) is a case study of the post-war military influence on basic research in the physical sciences and engineering at a single university, and offers further support for the thesis that military interest revolved around potentially or actually useful technique.

16. To be more concrete, the most specific expected military pay-off of HEP technique seems to have been production (rather than research) accelerators. See the discussion of the Materials Testing Accelerator and of 1950s work on particle beam weapons in Heilbron *et al.* (1981) and Seidel (forthcoming). Beam weapons are, of course, an integral part of the recently intensified intimacy of physics and the military in the US Strategic Defense Initiative.

17. The translocation of skills and material resources gained in a military environment to accelerator construction and experimental practice in HEP is documented in Heilbron *et al.* (1981) and Galison (1985). See also Galison (forthcoming). Here I am concerned especially with the transplantation of theoretical resources, on which see Schweber (forthcoming).

18. *Ibid.* See also Forman (1987) and, for interesting personal recollections, Bernstein (1987).

19. For more details, see Pickering (1984), pp. 60–73.

20. Schweber (1986) notes that the dominance of the pre-war US physics community by experimenters already fostered a pragmatic orientation in US theory. From this perspective, the significance of US military sponsorship of post-war HEP is that it promoted this distinctly American and pragmatic form of scientific life, theory included, to the forefront of international physics. The pragmatic US approach displaced the pre-war European style (which accorded much higher status to theory and theorists) as definitive of what physical research should look like.

21. Galison (1985, 1987). For more discussion of this point, see Pickering (forthcoming), and, for more recent developments, Pickering and Trower (1985) and Pickering, 'Constructing consensus: World views, institutional structures and policy-making in recent high-energy physics', paper presented at the 10th Annual Meeting of the Society for Social Studies of Science, Rensselaer Polytechnic Institute, 24–27 October 1985.

22. On the context of justification, see also *ibid.* Note that from the physicists' perspective the emphasis upon the production of theoretical knowledge as justification is unproblematic. It does, however, serve to obscure the military significance of the production of technique.

23. Cini (1980), p. 157: 'The dominant ideology in the U.S. lent itself particularly, through the mechanism of unbridled competitivity and the rat race, to the acceptance of a utilitarian and pragmatic, but fragmentary, concept of science with the consequent abandoning of its traditional aim of unification of knowledge.'

24. As already noted, much more exploration of these circuits is needed before the argument could be regarded as established. One significant area on which there is, as far as I am aware, remarkably little systematic information at present available concerns manpower flows, through HEP and into the military–industrial complex. The findings I have discussed here indicate that such flows were not, as one might otherwise have suspected, irrelevant to the constitution of scientific knowledge. Tracing their magnitudes and directions deserves high priority by historians of modern physics. It should also be of some concern to modern physicists.

11

Fermilab: Founding the First US 'Truly National Laboratory'

Catherine Westfall

In 1960 US physicists began to plan a multi-hundred million dollar accelerator laboratory. Although this laboratory followed a series of large accelerators sponsored by the federal government after World War II, it was the first designed to serve the entire national community of high-energy physicists. As planning for the first US 'Truly National Laboratory' proceeded, the US economy worsened and public scepticism about the value of science and technology intensified. In this environment, two questions emerged. Would cost and resulting pressures force changes in the way a major accelerator project was planned, organized and constructed? Would the federal government agree to fund a multi-hundred million dollar project even though the economic and social climate was less favourable to science funding than it had been in the 1950s? These questions were answered by the founding of Fermi National Accelerator Laboratory — or Fermilab, for short. This chapter examines the founding of Fermilab to chart the evolution of US accelerator development at a turning point in its history and probe the basis for the enduring US commitment to accelerator building.[1]

1 US Accelerator Building, 1945–60

In some ways, Fermilab was shaped by tradition. Plans for the expensive new machine arose from the momentum of accelerator building generated after World War II in the United States. This momentum was fuelled by a propitious funding environment, advancements in accelerator technology and a bounty of scientific achievements.

Science funding escalated rapidly with wartime programmes, most notably the successful atomic bomb and radar projects. Although attempts to continue funding started in World War I foundered because of lack of government interest and the scientific community's wariness of political control, during World War II the average annual federal research and development investment increased from less than $50 000 000 to $500 000 000.[2]

This wartime investment left rich new resources. In particular, the Manhattan Engineer District (MED), which administered the atomic bomb project, had undertaken massive construction. The MED built new facilities, including a laboratory to develop the bomb at Los Alamos, New Mexico, and provided new equipment and buildings for a number of existing academic institutions, most notably to the Berkeley Radiation Laboratory, where research was conducted on electromagnetic separation of uranium, and to the University of Chicago, where the first working pile was constructed.

The war also substantially increased the size of the American scientific community, since large numbers of scientists and engineers were trained for war projects. In addition, the nation had gained a number of talented scientists who emigrated to the United States to escape the political and religious persecution of Axis nations. Physicists, who had been culled primarily from academia, played a crucial role in wartime projects, and the investment in training and facilities for physical research was particularly heavy.

Since the wartime effort had been narrowly directed towards weapon research, investigation into the fundamental nature of matter had virtually halted. When this pressure lifted, however, physicists again turned their attention to the confusing and intriguing problems that had accumulated just before the war. They were particularly eager to study mesotrons, an unusual phenomenon spotted in 1937 in cosmic rays, which promised to provide information both about the atomic nucleus and about cosmic rays. This item on the research agenda provided a powerful stimulus for improving accelerators, the instrument designed for exploring subatomic matter. As in the past, physicists were eager to employ this powerful tool so that the desired phenomenon could be more copiously produced than in nature and studied under more controlled conditions. Since existing accelerators were unable to achieve or maintain the energies necessary for mesotron production, interest in devising a new, more powerful type of accelerator was high.[3]

Motivated by such interest, top scientists who had led war projects, such as Nobel laureates Ernest O. Lawrence and Isidor I. Rabi, were quick to exploit wartime contacts to obtain accelerator funding. Before the war ended, Lawrence obtained MED funds to support several accelerator projects at the Lawrence Radiation Laboratory (LRL), where he had built the first successful cyclotron in 1932. Support for the projects required a thirtyfold increase over the pre-war laboratory budget. Thanks to pre-war credentials and wartime contacts, Rabi, like Lawrence, was in a prime position to request funding. At his instigation, plans began in 1945 for Brookhaven National Laboratory, a major basic research centre to serve the many prominent universities in the north-eastern portion of the United States. Soon major accelerator projects were being planned at Brookhaven.[4]

While these plans were being made, some government leaders, such as Senator Harley M. Kilgore, felt that federal funds should be distributed geographically with an emphasis on fulfilling socially useful goals. None the less, large funding allocations were soon concentrated in a small number of national laboratories which had a generous allowance for basic research. Built from the foundation of wartime facilities, this national laboratory system was supported by the Atomic Energy Commission (AEC), established in 1946. Although the National Science Foundation was originally intended to be the primary sponsor of basic research, when it was finally established in 1950, its basic research budget was actually far smaller than that of the Atomic Energy Commission. For example, whereas the Atomic Energy Commission spent $26.1 million for physical research in 1949, the NSF began in 1950 with an appropriation of only $350 000 and the limitation that yearly budgets could not exceed $15 000 000.[5]

On the heels of the war, the idea of supporting accelerators through the atomic energy programme found support for many reasons. Because accelerators played an important role in the atomic bomb project, it was easy for leaders in government and the physics community to see them as a natural part of the nation's atomic energy arsenal and to feel confident that the public would support such a view. In addition, as leaders such as Lawrence and Rabi were quick to see, the atomic energy programme was sure to have a large budget, owing to post-war anxieties and hopes for the applications of nuclear energy. Although accelerators would primarily be used for basic, not applied, research, the devices were bound to become more expensive as they became more powerful; thus, it made sense to secure the funding of this important tool through a well-endowed agency.

Another factor that encouraged support for large accelerators through the atomic energy agency was a change in attitude. Owing to the success of wartime projects, government leaders, especially those in the military, were more willing to fund science than ever before. As Luis Alvarez, Lawrence's protege, has remarked: 'Right after the war we had a blank check from the military because we had been so successful.'[6]

Just as important, attitudes within the physics community had changed. Funding basic research through an atomic energy agency did not ensure the independence physicists considered crucial before the war. None the less, younger leaders, such as Rabi and Lawrence, emerged from the large-scale radar and atomic bomb projects less fearful of dependence on the federal government and more eager to grasp the opportunities made possible by government support, especially since exciting research questions that could best be explored with expensive accelerators had recently arisen.[7]

The AEC bound the physics community and the federal government into a new, compelling alliance. At a time when the promise of military applications of atomic research was a prime rationale for support of basic research, the agency provided ample basic research funding without military domination. At the same time, the federal government not only gained immediate expertise, but also provided for training the new team of experts needed for future programmes in advanced weapons and nuclear power. By the end of the 1950s another benefit appeared: working from World War II precedent, physicists promoted the view that accelerators were a natural part of the post-war atomic energy arsenal. In the midst of the Cold War, these instruments quite naturally joined space vehicles as symbols of technological, and therefore political, strength.[8]

All the while, science funding continued to increase. In 1950 government expenditures for research and development reached the all-time high of $1 billion, and by 1956 this amount had tripled. When the Russians launched Sputnik in October 1957, the pressure to upgrade science and technology prompted further escalation, so that from 1957 to 1961 federal funding for research and development saw a more than twofold increase. Even in an era of large overall funding increases, the record for particle physics was impressive. Whereas the subfield received $12.5 million from government agencies in 1950, the figure had jumped to almost $30 million by 1958.[9]

Ironically, although funding increases came from the desire to upgrade applied science and technology, in an era when physicists had considerable influence in funding allocation, particle physics flourished, although its prestige within the physics community derived from its ability to address fundamental issues — a characteristic which rendered it particularly remote from application. When justifying accelerator funding, physicists emphasized their traditional belief that basic research naturally led to useful applications, all the while assuming that government leaders, like physicists, also held the conviction that basic research deserved support primarily because of its intrinsic value. A propitious funding environment masked, but could not resolve, conflicting attitudes about the relative importance of basic and applied science.

At just the time when the new alliance with government made it possible to finance larger accelerators, technical innovations made it feasible to design such machines. Two advances crucial to the continued march to higher energies came from the élite groups of accelerator builders which grew at Berkeley and Brookhaven. In 1945 LRL researcher Edwin McMillan hit upon the concept of phase stability, an idea suggested earlier by Vladimir Veksler of the Soviet Union.[10] With the help of MED funding obtained by Lawrence at the end of World War II, McMillan began plans for a 300 MeV electron synchrotron, an accelerator which varied both the magnetic field and frequency to keep particles in phase. McMillan thus initiated a productive new line of synchronous accelerators capable of accelerating particles to much higher energies than previous machines. This innovation led to the Brookhaven Cosmotron, which accelerated protons to 2.2 GeV in 1952, and to the LRL Bevatron, which accelerated protons to 6.2 GeV in 1954.[11]

The Cosmotron and Bevatron had been built with the idea of exploring mesotrons, which were later identified as pions and muons, or, collectively, mesons. When the two accelerators began operation, however, they provided an unexpected dividend — they could create strange particles, the mysterious, long-lived particles first observed in the late 1940s in cosmic ray showers.[12] Without the control and intensity limitations faced by cosmic ray researchers, physicists at Berkeley and Brookhaven were able to make detailed, accurate measurements of strange particles. Numerous new particles were discovered. With the LRL Bevatron alone, researchers were able to discover dozens of particles in the next two decades. The flood of new data led to numerous exciting findings. For example, Chen-Ning Yang and Tsung Dao Lee discovered in 1956 that parity was not conserved in weak interactions.[13]

Although the growing army of small accelerators spread across the US produced a solid body of interesting information, a disproportionate amount of the research spoils went to Berkeley and Brookhaven, the chief US competitors in building large accelerators. Because of outstanding pre-war scientific credentials, Lawrence and Rabi were poised to assume power and influence during the war. After the war this power and influence translated into a competitive advantage for winning the most expensive fundamental research projects. These projects, in turn, brought new research success and prestige, which justified further special treatment. As physics funding continued to grow in the 1950s, the poor did not get poorer, but the rich got much, much richer, which created an imbalance that worsened as funding increased.

The first major reaction to this imbalance was the formation of the Midwest Universities Research Association (MURA) in 1954. At a time when Lawrence and Rabi championed the best interests of East and West Coast physicists interested in accelerator-based physics, Midwesterners were, in the words of one MURA researcher, 'left out in the cold'.[14] Determined to find a novel contribution to accelerator building and thereby gain entry into the AEC budget for accelerator building, MURA researchers investigated a number of ambitious schemes for a first-rate Midwestern accelerator. In addition to pioneering the concept of colliding beams, the particularly innovative MURA designers made a number of other contributions, including the application of the digital computer to accelerator design and advances in the understanding of the mechanics of particle beams.[15]

By the time MURA was formed, Brookhaven researchers had developed an innovation which allowed further development of conventional synchrotrons. In 1952 Brookhaven physicists expected a visit from European scientists planning a large machine for the newly formed European high-energy physics laboratory, CERN. In anticipation of the visit, Ernest D. Courant, Stanley Livingston and Hartland Snyder scrutinized Cosmotron designs to find ways to build a larger accelerator. In the process, they discovered the concept of strong focusing. (They later discovered that the idea had been proposed two years earlier in Greece by Nicholas Christofilos.) A strong-focusing, or alternating gradient (AG), machine has a series of strong magnetic lenses that focus horizontally, defocusing vertically, and then focus vertically, defocusing horizontally. The net focusing effect, which is much stronger than that obtained in previous accelerators, confined the beam to a smaller area, allowing a less expensive accelerator. Previously, it appeared that conventional synchrotrons more powerful than the Cosmotron and Bevatron would be prohibitively expensive.[16]

After the discovery of strong focusing, plans began for implementing strong focusing at Brookhaven, with the Alternating Gradient Synchrotron (AGS), and at CERN, with the Proton Synchrotron (PS). The discovery also stimulated MURA researchers, who devised a scheme for producing strong focusing which allowed the use of direct-current magnets and a much higher beam intensity. Such accelerators were dubbed Fixed Field Alternating Gradient Synchrotron (FFAG) machines. (MURA researchers later learned that researchers in Japan and the USSR had independently developed the same concept.) In the light of the scepticism about the feasibility of strong-focusing synchrotrons, it seemed quite possible that MURA was blazing a new trail. If conventional strong-focusing synchrotrons were unsuccessful, the MURA could well possess a more cost-effective and practical approach to future accelerators, through FFAG colliding beams and through higher-intensity, lower-energy machines.[17]

By 1960, however, the CERN PS and Brookhaven AGS had both passed their performance tests with flying colours. To the amazement of many physicists, these machines not only proved the feasibility of strong focusing, but also produced surprisingly high beam intensities. Almost immediately plans were drafted in the US for strong-focusing synchrotrons in the range 100–1000 GeV.[18]

By the time planning for the new accelerator began in 1960, a well-established tradition for building large accelerator projects had developed in the United States. An accepted élite of accelerator builders reigned at Berkeley and Brookhaven, where the nation's largest accelerator projects were designed, built and managed. Starting with the Brookhaven Cosmotron and the Berkeley Bevatron, major projects had alternated between the two laboratories. Since Brookhaven had built the AGS, Berkeley researchers expected that they would design, build and manage the next US high-energy accelerator.

With the increased complexity and expense of large accelerator projects, accelerator building at Berkeley had assumed a spirit of conservatism. Just before his death in 1958, Ernest Lawrence, who fervently promoted accelerator development in the 1930s and 1940s, worried that physicists would fail in their efforts to build the next high-energy accelerator. In line with Lawrence's attitude in the 1930s and 1940s, Edwin McMillan, Lawrence's successor, was eager to extend experimental capabilities and enthusiastically mounted a proposal for the new accelerator when the AGS proved the feasibility of strong focusing. None the less, Lawrence's later influence could be seen in the attitude of Berkeley designers, who prized reliability and eschewed risk-taking.[19]

While accelerator design at Berkeley assumed a spirit more conservative than that found in the 1930s and 1940s, attitudes towards outside users remained firmly grounded in the laboratory's past. Before World War II, when Lawrence obtained private funding, he and his students constructed the first small cyclotrons primarily for their own use. Although Radiation Laboratory accelerators were used from the beginning by researchers from other institutions, outside users were granted access to laboratory machines only by invitation, even after federal funding began in the late 1940s. In 1960 LRL researchers approached plans for the new laboratory both empowered and shackled by traditions developed in the preceding three decades.[20]

2 New Arrangements for Planning and Organizing the Accelerator, 1960–5

Once the feasibility of strong focusing had been proved, scaling up the design of a conventional strong-focusing synchrotron posed no major technical problem. Making non-technical plans for the new facility, however, was less straightforward. From the onset, even in corridors of tradition-conscious LRL, physicists agreed that the new laboratory, meant to serve the entire community of high-energy physicists, should be different from previous facilities.[21] Despite this consensus, physicists needed to define what would be different, a task that proved difficult against a backdrop of rising social and political tensions.

Part of the tension came from the fierce competition for designing the accelerator destined to be the prime US tool for particle physics for at least a decade. By 1961 the AEC was faced with three competing design teams: Berkeley's right to build the machine was contested both by Brookhaven and by a newly formed organization, the Western Accelerator Group (WAG).[22] WAG had been organized at the suggestion of Matthew Sands, an abrasive researcher from the California Institute of Technology (Caltech), who pioneered a scheme for cascading particles from one synchrotron to another more than a year before such work began at Berkeley or Brookhaven.[23]

When a meeting was called at the University of California at Los Angeles in late 1961 to lay ground rules for the design study, all involved, including Berkeley researchers, agreed to several tradition-breaking plans. Participants resolved that the accelerator would be designed in a study conducted at a central site with contributions from outside laboratories and that the design director should operate under the watchful eye of an advisory board with national representation. Further indication of change came in 1962, when AEC official Paul McDaniel declared that the accelerator would not necessarily be sited at a location near the design group, adding that the accelerator might best be managed separately from any existing laboratory.[24]

Despite such indications, competition for the proposal proceeded in a manner reminiscent of the previous decade. Despite their head start and subsequent innovative, economy-minded proposal, WAG researchers dropped from competition when it became clear that Berkeley, with its long experience in accelerator development, would dominate West Coast efforts. As was the case with the Bevatron/Cosmotron funding, the AEC, Brookhaven and Berkeley eventually agreed to a plan whereby both laboratories could build a new high-energy machine, one with higher energy than the other. Moreover, despite earlier public proclamations, in private McDaniel did not contradict Berkeley researchers when they said they intended to design, build and operate the new machine.[25]

In the meantime, the AEC and the physics community faced the problem of accommodating the expensive programme in the already swollen high-energy physics budget. This problem was particularly acute because MURA's FFAG, now estimated at $148 million, was slated for authorization. In the light of the unexpectedly high beam intensities of the AGS and PS and the greater research potential of high-energy machines, it seemed that the FFAG might be cancelled, a decision sure to embitter already antagonized MURA researchers. For help in adjudicating accelerator plans, the AEC convened a panel of eminent scientists chaired by Norman Ramsey.[26]

The most controversial issue facing the Ramsey Panel was the relative importance of the high-intensity accelerators, most notably MURA's FFAG. In its final report, the Panel noted that an increase in proton energy was 'the single most important energy parameter to be extended'. In addition to clarification of the weak and strong interactions, the Panel hoped that further high-energy research would bring physicists closer to understanding the fundamental relationship of all matter by providing 'a clue to the connections among the different kinds of basic force'. The Panel admitted that high-intensity beams allowed researchers 'to study, in detail, interactions of such low probability that they have not yet been observed using present accelerators'. None the less, the report mentioned technological problems associated with high-intensity accelerators — in particular, the complications caused by high radioactivity levels — and noted the 'relatively high cost of high intensity machines'.[27]

On 26 April the Ramsey Panel issued its report. Their first recommendation was that the government: 'Authorize at the earliest possible date, the construction by the Lawrence Radiation Laboratory, of a high-energy proton accelerator at approximately 200 BeV energy.' They also recommended that Brookhaven be authorized to construct the proposed AGS storage rings and begin 'intensive design studies' for an accelerator in the 600–1000 GeV range. All large laboratories, including Berkeley and Brookhaven, 'should incorporate an administrative structure with national representation to assure that all proposals for qualified scientists shall be considered on equal footings'. Despite the vociferous pleas of frustrated MURA physicists, who had been working for more than ten years to fulfil the dream of a first-rate Midwestern accelerator, the Panel's support for the FFAG was conditional. In its final form, the Panel's specific recommendation was that the machine should be authorized 'without permitting this to delay the steps toward higher energy'.

Ramsey later admitted that the Panel's recommendation was really the 'kiss of death' for MURA.[28]

Although MURA's fate remained unclear in spring 1963, in line with the recommendations of the Ramsey Panel, LRL received permission in April to begin a detailed design, and the 200 BeV project was launched. Concurrently, Brookhaven was authorized to begin studies of a machine in the 600–1000 GeV range. Although Berkeley researchers knew that McMillan would work with an advisory committee to solve management issues, they hoped that, for all the talk of the changes necessary for the large new facility, very little would change. At this point they had every reason to believe that LRL scientists would eventually exercise the traditional prerogative of accelerator designers and build and manage the new machine. LRL, with its reputation of past and current excellence, had once again ushered in plans for designing a new, more powerful accelerator.[29]

Despite such hopes, in 1964 a series of mutually reinforcing pressures and resulting social tensions changed both management and siting plans for the new accelerator. In January MURA's fate was sealed. Owing to the tightening budget, the emergence of expensive Berkeley and Brookhaven proposals and the lack of enthusiastic support from the physics community, the MURA proposal was cut from the budget.[30]

The MURA defeat intensified two complaints which had their roots in the previous two decades. In line with Senator Harley M. Kilgore's push for geographical distribution of research funding in the late 1940s, Midwestern politicians campaigned against the inequitable geographical distribution of research funding, taking on MURA as a *cause célèbre*. At the same time, the wrath of Midwestern scientists, which had been building since the 1950s, exploded. Once again, the Midwest had been passed over in favour of already prominent institutions. Although Midwestern scientists argued for fair funding distribution on the basis of meritocracy, while Midwestern politicians argued for the evenly geographical distribution of funding, the combined force of both groups produced considerable political pressure for a Midwestern accelerator, undermining Berkeley's plans to build the 200 BeV accelerator in California.[31]

Berkeley plans were further undermined by new tensions arising from the unprecedented expense of the new accelerator. Because of its high cost, physicists realized that only one such accelerator could be built. In the light of this realization, young active experimentalists, such as Leon Lederman, the current director of Fermilab, were quick to raise the banner of outside-user rights. In 1963 Lederman, a member of the 200 BeV Advisory Committee, wrote an informal report for a 1963 Brookhaven summer study, entitled 'The Truly National Laboratory (TNL)', using the initials TNL as a pun on BNL, which he felt was not functioning as a truly national facility. Lederman noted that co-operation and enthusiasm for the new facility would only be assured 'when it is clear that the new facilities are accessible *as a right* to any physicist bearing a competitively acceptable proposal'. The ideal laboratory he described had complete on-site facilities for outside users, resources for facilitating individual experiments, an accessible and pleasant site, scheduling and advisory committees to assure fairness in the allocation of beamtime and free communication between management and users, and a strong laboratory director 'responsible to a governing body of wide national representation'.[32]

Although some, like Lederman, complained that Brookhaven did not offer fair access to outside users, for many experimentalists, the likelihood of LRL control of the new accelerator prompted even more concern. In line with Lederman's suggested management scheme, the Advisory Committee subsequently advocated management by a corporation with nationwide representation. Although this plan, which would abolish the traditional prerogative of accelerator designers to manage the machine, had strong support from outside-user advocates, LRL leaders vehemently opposed any major change from tradition. As McMillan noted in an April 1964 letter to a colleague, LRL researchers wanted 'to continue the line of development started many years ago by Professor Lawrence'. Although McMillan agreed that the new laboratory should be a national facility, he stressed that an accelerator is not 'an industrial plant, to be located where economics dictate and operated by hired hands from the region and managers exported from the home office'. In McMillan's view, forming a nationwide management group would be both difficult and time-consuming.[33]

Despite McMillan's objections, the impetus for national corporation management was further fuelled by concern for the funding prospects of the expensive new machine. Conflicting views about the relative value of basic and applied research, previously masked by ample funding, now emerged. As Figure 11.1 shows, at just this time the proportion of basic research funding

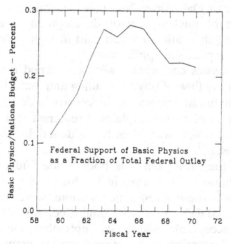

Figure 11.1   Federal obligation for basic physics research as a fraction of the total federal budget. Data taken from Physics Survey Committee, National Research Council, *Physics in Perspective* (Washington, 1972), p. 648.

was rising relative to overall federal expenditures. Whereas basic research consumed just over 0.1 per cent of federal expenditures in the late 1950s, by 1965 it commanded about 0.25 per cent. High-energy physics, which, unlike other subfields, received almost all its funding (90 per cent) for the Atomic Energy Commission, commanded a very noticeable portion of the basic research budget for that agency. Moreover, since accelerator projects required large chunks of construction funds, they formed an especially visible portion of the high-energy physics budget.[34]

The highly visible chunks of money requested for high-energy physics research prompted complaint both inside and outside the physics community. In March 1964 the AEC held budget hearings with the Joint Committee for Atomic Energy (JCAE), the powerful congressional overseer of the AEC and traditional champion of atomic energy programmes, including accelerator projects. Chester Holifield, JCAE chairman and Representative from California, pointed out that the high-energy physics budget had jumped from $53 million in 1960 to $135 million in 1964. He was concerned that 'the overall budget of the AEC stays about the same, but the high-energy part of it is increasing . . .'. Conceding that it was 'perfectly natural' for scientists to have a limitless supply of ideas and ambition for exploring them, he none the less worried that in accommodating such researchers, decision-makers were 'squeezing to death many fields of science'. 'Very frankly', he concluded, 'the Congress is becoming alarmed . . .'.[35]

Proponents of the development of nuclear reactors were also upset. At the hearing Holifield quoted statements by Eugene Wigner. 'As a result of the economy measures', Wigner explained, 'there is a serious shortage of funds in other areas, and even low-energy nuclear physics suffers heavily.' Wigner was not the only one worried about the situation. For several years, Alvin Weinberg, Director of Oak Ridge National Laboratory and, like Wigner, a long-time advocate of nuclear reactor development, had questioned the wisdom of increasing the national investment in high-energy physics, noting that the field had few practical applications.[36]

Leaders of the physics community were concerned that such complaints would slow the healthy flow of physics funding and make it difficult to obtain the large, noticeable funding allocations necessary to design and construct the accelerator. In addition, such complaint threatened to further splinter the physics community, which was already burdened by the unhappiness of Midwestern physicists.[37]

In the wake of increasing tensions both inside and outside the physics community, in autumn 1964 all eyes in the high-energy physics community turned to Berkeley, where intense negotiations were under way between McMillan and the 200 BeV Advisory Committee. Amid numerous attempts to find a mutually acceptable plan in late September, the Advisory Committee suggested the formation of a national management corporation. This corporation would join with the Regents of the University of California in a 'Joint Venture', which would supervise the design and construction of the new accelerator. Although the University of California was allowed a vote, along with the AEC, in the selection of the new laboratory Director, the laboratory would 'become the sole responsibility of the National Corporation' after construction of the machine. Despite McMillan's vehement objections, the Advisory Committee retained the Joint Venture scheme in their final recommendation, which it sent to him on 11 November 1964.[38]

With the Advisory Committee report in hand, McMillan had the next move. Although the Committee's plan for national corporation management was responsive to outside-user pressures and the need to gain unanimous support from the physics community, McMillan faced a variety of pressures to resist the non-traditional management scheme. For one thing, he faced the expectations of his staff. As explained in a 13 October memo to Edward Lofgren, who headed the Berkeley design team, LRL designers felt that new plans forced an ill-advised distinction between machine builders and users. As design team member Glen Lambertson recently recalled, the team was horrified by the unprecedented idea of building the complicated machine and turning it over at some arbitrary point to another group, since accelerators often require an unpredictable amount of fine tuning. As Lambertson also admitted, part of the problem was that designers had joined the effort expecting the traditional prerogatives of machine builders. McMillan also faced enormous resistance to change from Berkeley's star team of active inside users, who were just now struggling to adjust to outside-user representation in the scheduling of beamtime at the Bevatron. Even more important, McMillan saw the push for non-LRL management as an unworkable idea prompted by jealousy and opportunism. At this point, the pressing desire to continue Lawrence's tradition was a great burden.[39]

McMillan wrote to McDaniel on 23 November 1964, noting that a national management corporation 'offers a valid mechanism for assuring national representation, and in the present context probably the best way'. None the less, he urged that the matter of turning over eventual operation of the machine to the corporation 'be kept open for further discussion', leaving management arrangements unsettled. At this point management plans were commandeered by NAS President Frederick Seitz, who was determined to see the rapid formation of a new management organization to gain support among physicists for the new facility. Seitz first conferred with AEC chairman Glenn Seaborg and approached representatives from Brookhaven's management organization, AUI, who decided not to take on the project. Empowered by his considerable influence as NAS President, Seitz took the decisive step in late November: he called an informal NAS meeting and plans were made for inviting 25 university presidents to a January meeting to cast future plans for large-accelerator projects.[40]

This step, which also led to the opening of the site competition, was responsive to pressures both inside and outside the physics community. The new organization kept hopes alive for a Midwestern accelerator, provided assurance for outside-user advocates, and promised to unify the high-energy physics community and optimize the chances for continued funding. Although it was modelled after AUI, Brookhaven's management organization, the new organization played a new role. AUI had been formed before the first large-accelerator proposal was mounted at Brookhaven, and thus Brookhaven accelerator designers, like those at Berkeley, had built and managed the accelerators they had designed. In contrast, the new organization threatened to usurp the traditional prerogatives of Berkeley accelerator designers. The cost of the 200 BeV, the tightening budget and resulting social tensions had forged new plans. For the first time, a group of accelerator designers had no assurance that they would be allowed to manage and build a machine at the site of their choice.

3 Siting, Funding and Constructing the Accelerator, 1965–72

From 1965 to 1972, in the midst of a worsening economy and strident public questioning of the value of science and technology, the final steps were taken in the founding of the first US 'Truly National Laboratory': a site was chosen, the first large funding allocation was obtained and the accelerator was built. By June 1972, when the new laboratory was officially named Fermilab, a new course had been set for planning, organizing and construction of major US accelerator projects.

From the onset, AEC Chairman Glenn Seaborg realized that site selection would be politically sensitive. To preserve overall AEC authority while protecting the Commission from the repercussions of becoming the sole arbiter of the competition, Seaborg requested help from NAS President Seitz. The NAS subsequently entered into a contract with the AEC, agreeing that a NAS Site Evaluation Committee composed of eminent scientists would evaluate all candidates and present a short list of finalist sites to the AEC Commissioners, who would choose the final site.[41]

Despite these careful plans, site selection was time-consuming and difficult. After the competition opened in April 1965, the AEC was deluged with proposals. Berkeley submitted two California sites for consideration and a group of Midwestern physicists spearheaded a proposal for Madison, Wisconsin, the site of the former MURA headquarters. In the wake of speculation that the 600–1000 GeV machine might never be built because of the tightening budget, Brookhaven submitted a proposal. Since many felt that the accelerator would bring prestige and economic benefit to its community, a large number of proposals were also submitted by citizen groups with no affiliation to the high-energy physics community or to the AEC.[42] In the end, the AEC was faced with the task of evaluating 200 sites in 46 states. When the Commission tried to skim about 35 sites for referral to the NAS Committee, the President himself intervened. President Johnson, a Texan, insisted that a site in Austin, Texas, be restored to the list.[43]

To satisfy Johnson, extricate the AEC from the uncomfortable position of eliminating sites with minimal requirements and increase the base of congressional support, the AEC subsequently forwarded 85 sites to the NAS Committee. While this move strengthened the position of the AEC and increased funding prospects for the accelerator, it placed a considerable burden on the Committee. In addition to the increased workload caused by the large number of sites, the Committee simultaneously faced judging all sites with minimum requirements and the growing realization that site selection would hinge on intangible factors. At this point it was clear that site selection would be a highly politicized, public spectacle, a turn of events that made many physicists uncomfortable, since such decisions had always before been made within the confines of the physics community. As one Committee member wrote to Chairman Emanuel Piore, because of the political implications of the contest, both inside and outside the physics community, it would be 'foolish to think that our only job is to come up with the best possible site. . . . Unless it is quite universally accepted and supported our exercise will have been a fruitless one.' As with management and design plans, cost and resulting pressures made non-technical issues dominant in the site contest.[44]

As planned, the Site Committee made recommendations for finalist sites in a March 1966 report. The report explained that although none of the sites was 'ideal', the Committee found several sites which 'in general' satisfied all important requirements. Brookhaven and one of the California sites, the Sierra Nevada Site, made the list, although the California Camp Parks Site was discounted because of the risk of earthquakes. The list also included sites in Ann Arbor, Michigan; Denver, Colorado; Madison, Wisconsin; and two sites in Illinois — South Barrington or Weston. Although the Denver site was not associated with past tensions, the site contest included sites sponsored by the two traditional rivals — Berkeley and Brookhaven — and several Midwestern sites at a time when many felt that the region was long overdue for a major accelerator. The AEC faced a difficult decision.[45]

Site selection was only one of Seaborg's problems, however. As the contest continued, the Chairman became increasingly concerned about funding prospects for the project because of the declining federal budget. Burdened with the war in Vietnam and burgeoning Great Society programmes, Johnson began to stress austerity in mid-1965. Throughout the year the economic situation worsened. By December White House advisors warned that unless spending drastically declined, a hefty tax increase would be needed to avoid inflation, a step Johnson was loath to take. This pressure, along with emerging questions about the social value of basic research, created an economic and social climate much less favourable to accelerator funding than that enjoyed by physicists in the previous decade.[46]

Funding difficulties for the new accelerator were compounded by tensions within the physics community, where divisiveness spurred by the MURA defeat was exacerbated not only by the site competition, but also by a flurry of criticism that erupted when LRL designers released their design in June 1965. The most vociferous complaint came from Cornell physicist Robert Wilson, who produced a series of critical letters and reports that circulated widely throughout the physics community and the AEC in the autumn and winter of 1965. The criticism began with a September 1965 letter to McMillan. Although the details were 'most professionally worked out', Wilson remarked, the design was 'much too conservative', and 'lacking in imagination'. Wilson added that he was 'offended' by the $348 million Berkeley cost estimate, which was 'ridiculously high'. Wilson produced an alternative design, estimating that the components of his machine could be built for only $70 million. Worried that criticism would delay the project, Seaborg and other leaders organized two meetings to air design disagreements within the physics community. No consensus for redesigning the project emerged, however, and design criticism died down.[47]

Soon, however, the project was delayed by a new crisis in the site selection contest. The Commission had hoped to move quickly after the NAS Site Committee announcement of finalist sites in March 1966. Optimistic estimates indicated that the final site might be chosen as early as July. By this time, however, Clarence Mitchell of the NAACP complained of the lack of open housing legislation in Illinois, prompting new, time-consuming investigations into the status of civil rights in all states containing finalist sites.[48]

Despite civil rights complaints, the AEC announced on 16 December 1966 that the accelerator would be built in Weston, Illinois. The Commission was immediately besieged with opposition. LRL physicists refused to endorse the site and Mitchell decried the choice because of the lack of fair housing in Illinois. Ironically, in the light of NAACP complaints, others claimed that the site was chosen as a result of a political deal President Johnson made with Senator Everett Dirksen, who suddenly changed from vehement opposition to support of an open-housing bill. Although such accusations continue, close inspection of the written record and other evidence suggests that the Commissioners sincerely considered the official site criteria and chose the site themselves. While the Commissioners were undoubtedly pleased to choose an Illinois site because of previous Midwestern complaints, the decision apparently hinged on the site's accessibility to outside users.[49]

By the time the site was chosen, the new management organization, the Universities Research Association (URA), had evolved into a functioning organization. Within days of the site announcement, URA President Norman Ramsey negotiated a temporary contract with the AEC and organized meetings to choose leaders for the new laboratory.[50]

URA's task was complicated when Edward Lofgren declined the job of Design Director. In addition to feeling dissatisfied with his job description, which seemed to preclude directing the laboratory, Lofgren felt unsure of the Weston site and unconvinced that he could assemble the necessary staff 'and develop an organization having the enthusiasm and spirit needed to make the projects a distinguished success'. After some deliberation, URA decided to offer Wilson the position of Laboratory Director. While Wilson had never built one of the large proton synchrotrons, he was a well-respected physicist, had directed the laboratory at Cornell for twenty years, and had recently directed the construction of an electron synchrotron at Cornell, which had come in on budget and ahead of schedule. Ironically, although Seaborg, Ramsey and other leaders had been annoyed by Wilson's economizing schemes in 1965 because they seemed to threaten the project, Wilson's stress on economy was now a point in his favour, owing to budget limitations.[51]

Although Wilson's appointment was endorsed by the AEC and many outside the Berkeley circle, most LRL physicists were enraged with the choice. LRL designers were also convinced that the accelerator would be prohibitively expensive to build at Weston. Chester Holifield and Craig Hosmer, the two California JCAE members, took up the cause of the despairing LRL group, creating further problems for the AEC on the eve of the authorization hearings. Seaborg was understandably concerned, since the project had little chance for survival unless the JCAE recommended the laboratory's first large funding allotment, an allotment that would signal a federal commitment to complete the project, due to the size of the investment.[52]

With the tensions prompted by the MURA defeat, the site competition, disagreements about the Berkeley design and the tightening budget, it was clear by the time of authorization hearings that the 200 BeV project had suffered from a variety of disadvantageous social and economic pressures. At the same time, however, the project enjoyed a number of benefits, which drew largely from the considerable capital of scientific interest and prestige accrued by high-energy physics research. The accomplishments of the field were praised directly in promotion efforts by the AEC and the physics community. In addition, the prestige and interest inspired by the subfield bolstered the project's prospects in more subtle ways. The project benefited from URA efforts to quell site and design disagreements within the physics community, efforts that relied heavily on the often-repeated admonition that without unanimous support the community would not get the machine, an admonition that only worked because large numbers of physicists strongly wanted the new research tool.[53]

The project also benefited from the strong advocacy of Seaborg, whose support grew largely from a commitment to sponsoring basic research, support which was undoubtedly intensified because of the high prestige of the field. The great public interest in the site selection contest also stemmed from public acceptance of the main tenets of those promoting high-energy physics: the prestige of the field and the promise that practical benefit would accompany further investment in accelerators. Although the fervour of the battle complicated the task of the Commissioners and increased tension within the physics community, it also enhanced funding prospects by widening political support for the project and turning attention from the issue of whether the project should be funded to the issue of where it should be sited.[54]

At the authorization hearings the AEC valiantly promoted the virtues of the project and defended the controversial choice of Weston. In the face of budgetary pressure, the subject of a low-budget expandable machine arose. Although LRL designers had considered ideas for expandability since 1964 and were about to publish a well-honed expandability scheme, the idea now became associated with Wilson, since the Commissioners were stressing his willingness to apply innovative ideas within a budget. The attractive combination of an expandable machine and an energetic, economy-conscious director helped secure support, even from the California Congressmen who were still disgruntled because Berkeley lost the machine.[55]

After the JCAE recommended funding in April 1967, the funding bill was vigorously debated in the Senate. In the midst of this debate, a *New York Times* editorial angrily articulated the mood of many Americans in 1967, the year that race riots erupted in Newark and Detroit and anti-war protesters marched on the Pentagon. 'The nation is engaged in a bloody war in Vietnam; the streets of its cities are swept by riots born of anger over racial and economic inequities', the editorial cried. 'It is a distortion of national priorities to commit many millions now to this interesting but unnecessary scientific luxury.' Such scepticism about the relative social value of science was destined to grow. Despite complaints, however, Congress eventually passed the funding bill. On 26 July 1967 Johnson signed the bill and the first large funding allotment for the 200 BeV project was secured. As the funding of the new laboratory proved, the US Government was willing to continue the post-war tradition of building ever-larger accelerators, even though the economic and social climate had become much less favourable for science funding.[56]

By the time the funding bill was passed, plans for the new laboratory, which Wilson temporarily named the National Accelerator Laboratory (NAL), were already under way. Since the URA made provisions for a strong director, Wilson immediately assumed a dominant role in defining how the new laboratory would be established. To accomplish this task, the new Director had $243 million at his disposal, since the Berkeley design team had spent $7 million of the $250 million total budget. Many physicists questioned the feasibility of building a viable 200 GeV accelerator within such a budget; building a viable 200 GeV expandable machine seemed even more unlikely. Wilson was undaunted by such scepticism. Owing to economic pressures and his innate enthusiasm for producing maximum capability within strict budget limitations, Wilson was happy to accept the challenge of building a 200 GeV expandable machine for $243 million.[57]

The design would have been reconsidered at this point in any event, to take individual site factors into consideration and refine other elements of the design. However, considering the already apparent conflicts between the Wilson and LRL approaches to accelerator design and the fact that a new team would construct the machine, it was clear that the 200 BeV design would be revised extensively. The task was also complicated by Wilson's determination to present costs and schedules to the AEC in time for the 1968 budget.

To accomplish this goal, Wilson gathered a large number of experts, including members of the LRL design team, in the Chicago suburb of Oak Brook in June 1967 and set a mid-October deadline for a preliminary design.[58]

Wilson later explained that he approached the design much as he approaches a sculpture. From the beginning he had a rough idea of an accelerator capable of much higher energy, but as various ideas were suggested, he tried to keep his image of the machine fluid so that innovative ideas could be incorporated. He was also intent upon keeping a balanced view of the whole machine in his mind.[59]

Wilson's concern for balance also influenced the way he approached other aspects of the design study. In his opinion, as Laboratory Director he needed to keep a good balance between the interests of the physicists, who would use the accelerator, and the interests of the federal government, who would pay for the accelerator. If the machine could not be used for exciting physics experiments, physicists wouldn't use it. At the same time, if the machine was too expensive, the government would refuse to pay for it. Wilson feels that this approach to accelerator design resembles that of his former teacher, Ernest Lawrence. As Wilson explains, 'Lawrence designed machines based on the amount of money he had available. . . . Once he figured what he wanted to build, he expected little difficulties, and then refused to back down from his goal, ever.' For Wilson, Lawrence's lesson was that the accelerator builder should strive for optimum physics capability and minimal cost. Ironically, this lesson, which incorporated Wilson's characteristic desire to achieve optimum benefit from available resources, led him to take risks unacceptable to Edwin McMillan, the physicist Lawrence chose to succeed him. While the Berkeley team concentrated on producing a conservative design that would be sure to work right away, Wilson insisted that the Oak Brook designers take risks, believing that whatever failed could be fixed later. In his opinion, if the accelerator worked immediately, it had been overdesigned.[60]

Although Wilson saw the design effort as an adventure, some at the summer study were dismayed at his design style, which they saw as flamboyant and risky. A particularly vivid reflection of Wilson's style could be seen in the design of the main ring magnets. In line with his 1965 suggestions, Wilson promoted the design of compact, H-shaped magnets. Instead of the combined-function magnets shown in the Berkeley design, Wilson decided to use separate magnets for bending and focusing, an idea conceived by Gordon Danby. Although separated-function magnets had never before been used in a proton synchrotron, they promised more flexibility in accelerator operation and allowed more efficient beam focusing. In addition, while Berkeley schemes for an expandable machine hinged on the later addition of magnets, by using separated-function magnets, Wilson was able to design magnets that could go to energies as high as 500 GeV with the later addition of power supplies. The resulting design took the risk of placing magnet coils much closer to the beam than had been the case in previous designs.[61]

The design was cleverly innovative: in addition to the advantages of separated-function magnets, the compact magnets required smaller tunnels and less steel, and thus saved money. In addition, Wilson's scheme for expansion could be implemented without the lengthy down-time required by Berkeley schemes. None the less, critics felt that the risk of using untried features, compact magnets and magnet coils close to the beam was unacceptable, considering the cost and importance of the project. Some physicists also criticized the economical, but inconvenient, main-ring tunnel, which was small and lacked cranes or air conditioning, and the decision to omit caissons as foundations for the main ring magnets.[62]

By October, Wilson had a preliminary design. Although some accelerator experts went so far as to call it 'irresponsible', Wilson obtained approval for his ideas from both the AEC and the URA. After receiving this approval, Wilson gave the AEC a brief summary of the design in mid-October, as planned. This summary itemized costs totalling $243 million and promised that the machine would be in operation by June 1972. Although criticism resurfaced periodically, particularly during times of trouble, Wilson had won the right to build the machine his way.[63]

In the next two months Wilson and his staff produced a full-scale design report describing the design in more detail. When the report appeared in early January, it described a machine that in some ways resembled that shown in the 1965 Berkeley design. Like the Berkeley design, the NAL design featured a four-stage machine in which protons were accelerated by a Cockcroft–Walton accelerator, transferred to a linear accelerator for more acceleration, transferred to a booster synchrotron for further acceleration, and finally transferred to a large synchrotron for acceleration to design energy. In addition to this cascade design, the NAL design incorporated the long straight sections suggested by Thomas Collins, as in the LRL design.[64]

The NAL report differed from the LRL report in that it presented additional innovations, such as the use of separated-function main-ring magnets and a new beam-extraction design. Unlike beam extraction in the LRL report, in the NAL report the beam for experiments emerged from the accelerator at only one point — a scheme made feasible by a clever new septum for extracting the beam, developed by former Brookhaven researcher Alfred Maschke. In addition to saving money, the scheme was simpler and therefore more reliable.[65]

However, the most striking differences between the two designs stemmed from Wilson's insistence upon achieving maximum capability for minimal cost. The original Berkeley design featured a 200 GeV accelerator laboratory costing $348 million. Although the Berkeley group later produced schemes for a reduced-scope machine that would cost about $20 million less than the total NAL budget, such schemes produced beams that were much less intense. In contrast, the NAL design featured a $248 million 200 GeV accelerator laboratory with a machine that was expandable to as much as 500 GeV and produced a beam slightly more intense than that in the original

Berkeley design.[66]

Comparison of the designs reveals how Wilson was able to achieve a machine with greater capability for less money. For example, one saving came from minimal magnet costs, which comprise a substantial portion of the total cost of the accelerator. While the Berkeley design required more than 17 tons of steel for the main-ring magnets, the NAL design required less than 9 tons. Figure 11.2 gives additional pictorial evidence of the difference between the two designs, showing the LRL magnet design in the lower part of the figure and the NAL magnet design in the top figure. Note the more compact design of the magnet, the smaller tunnel size and the absence of the crane shown in the NAL design. In the LRL design also note the substantial foundation used in the Camp Parks site, which was used for cost estimating. Ramsey explained later that Wilson took risks on about twenty aspects of the design, saving an average of about $5 million for each risk. 'We knew something would fail,' he noted 'but we figured it would be much less expensive to fix the failure than to play it safe with all 20 items.' As one of the machine's designers recently concluded when looking at NAL magnet plans: 'It was a design pushed to the limit.'[67]

To finish the job within the promised deadline and budget, Wilson and his deputy, Edwin Goldwasser, employed a number of distinctive schemes. For example, Goldwasser remembers that 'instead of giving the production of magnet laminations to one contractor, we gave one-third to one and a second one-third to another, telling them the contract for the last one-third would be awarded to the competitor who finished faster and cheaper'. Wilson also encouraged spartan standards when constructing the machine and used competition to spur the progress of his staff.[68]

As he later reported, as a result of such schemes, Wilson was quite 'confident' about the laboratory's prospects as mid-1971 approached. Construction progress was extremely encouraging. As he explained:

> Both our Linac and Booster had been brought into operation [at full design energy] before ... July [1971]. Our main-ring magnets were all in place in the tunnel and all were connected to the power supply. The radiofrequency devices which do the actual pushing of protons were also in place. The computer controls were ready. Indeed we were able to inject protons into the main ring ... even make them go around the 4-mile-long trip several times.[69]

Fuelled by this confidence, Wilson told both the JCAE and the NAL Users Group in spring that the laboratory might produce a 200 GeV beam in mid-1971.[70]

Although NAL and URA leaders had expected some aspects of the accelerator to fail, in line with Wilson's design style, they were not prepared for the traumatic summer and autumn of 1971. The main ring contained more than 1000 magnets. When researchers tried to bring them into operation they found, to their horror, that a high percentage failed. This was frightening

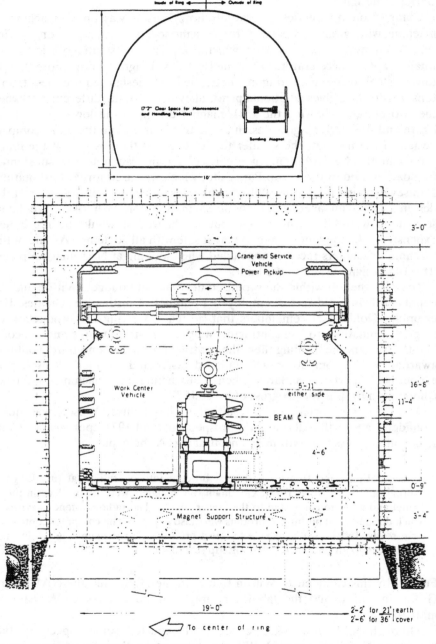

Figure 11.2 Comparison of 1968 National Accelerator Laboratory design (top) and 1965 Berkeley design (bottom). Note difference in tunnel size. From National Accelerator Laboratory, 'Design Report' (Batavia); '200 BeV Accelerator Design Study' (Berkeley), Vol. 2, UCRL, p. vi–11.

because of the number and expense of the magnets, and because researchers knew the entire project would fail unless the problem was solved. After several months of testing, one major source of trouble was discovered. As the 1971 URA Annual Report later explained: 'With the high summer humidity the magnets in the cold tunnel became saturated with moisture and any small cracks in the epoxy-impregnated fiberglass provided a conduction path to ground that could lead to a serious spark discharge.' At the time Wilson felt that a contributing factor to the coil design problem was the lack of air conditioning in the tunnel, one of the controversial cost-saving measures that emerged from the Oak Brook summer study. As he later told the JCAE, the $1 million cost of air conditioning was 'clearly a mistake'. Steps were subsequently taken to heat and dry the tunnel, methods for reducing cracks were devised and more thorough initial tests were developed. By the beginning of 1972, magnet failures had declined and technicians had learned how to replace magnets quickly when they did fail. However, as Wilson admitted to the JCAE, by that time approximately 350 magnets had failed, 6 months had been lost and about 10 per cent of the original cost of the magnets had been spent to overcome the difficulties.[71]

Although main-ring magnets caused the most worry, other problems also plagued the NAL staff during this period. As the 1971 URA report explained, main-ring problems 'were frustrated by linac and booster failures' which continued until late 1971, 'since beam studies of the main ring could be undertaken only when the linac and booster were operating well'. Another problem was caused by obstacles in the main-ring vacuum tube. As Wilson recently explained, 'there were little shavings of stainless steel from cutting and welding the magnets' during magnet replacement. To solve the problem, researchers first tried to train a ferret, affectionately named Felicia, to drag a harness through the tube to collect the pieces. When Felicia refused to go through some areas, an ingenious mechanical device was developed to replace her.[72]

The URA Annual Report, which was issued on 18 January 1972, noted that the previous year 'had been a difficult one at the National Accelerator Laboratory'. Although staff members first rallied to solve problems, many found the continuing main-ring problems exhausting and demoralizing. As Helen Edwards, who worked on the problem, explains: 'The whole period is a kind of blur. All I remember is infinite trips around the main ring to find out which of the magnets had failed.' The resulting stress also exacerbated tensions between Wilson and his staff. In line with the feelings of some at the 1967 summer study, some staff members felt oppressed by Wilson's leadership style, which they saw as impulsive and domineering. One participant remembers that at the height of the magnet crisis someone half-jokingly passed around a copy of Albert Speer's memoirs so that staff members could note the similarities between Adolf Hitler and their leader.[73]

The magnet problems also caused trouble outside the laboratory. Many users were terribly annoyed at the delay, since they had hurried to prepare experimental equipment based on Wilson's optimistic projections. In addition, the problems gave credence to arguments against Wilson's design style. Wilson and Goldwasser remember hearing rumours that a campaign was being mounted to remove Wilson from the directorship. Outside pressure from irate users and opposition to Wilson, both inside and outside the laboratory, further eroded the morale of those desperately trying to fix main-ring problems. As Edwards explains: 'It was a grim time.'[74]

After devising a new, non-hierarchical organizational plan which focused effort on solving main-ring magnet problems, Wilson began to see steady progress by late 1971. After intensive beam studies in December 1971, researchers were able to produce a 20 GeV beam on 22 January 1972. In addition to the encouraging increase in energy, researchers were happy to note that the beam seemed stable from pulse to pulse, indicating 'that the injector, magnets and magnet power supply, and radiofrequency acceleration system all work properly together'. In addition, the magnets seemed to be performing well. Improvement continued. On 11 February a 100 GeV beam was obtained, breaking the world's record for proton energy, held by the USSR. After steady increases in energy throughout the month, 200 GeV was achieved for the first time on the afternoon of 1 March 1972. As it happened, JCAE authorization hearings were in progress at the time.[75]

J. Richie Orr, who held an important administrative post at the time, remembers that someone sarcastically commented that Wilson's efforts in early 1972 were aimed at 'trying to set the indoor proton speed record'. To get the first 200 GeV beam, several modifications were made, including schemes for keeping the magnets cool, since the water-cooling system had not yet been installed. None the less, reaching 200 GeV was an important milestone, for it sent a message to physicists, both inside and outside the laboratory, and to government leaders that the project would be a success. After reaching 200 GeV, Wilson wasted no time in pushing to the next goal. In July 300 GeV was obtained, and by the end of the year the accelerator reached 400 GeV. When the accelerator reached 100 GeV in February, a Soviet–US collaboration started the first experiment, which measured the size variation of the proton at different energies of collision.[76]

Although it would take some time before beam was available to the experimental areas, and the laboratory struggled throughout 1972 to raise the intensity to the promised level, Wilson and his staff had reason to celebrate by the June 1972 deadline. The project had come in ahead of schedule and about $20 million below budget, and an exciting physics experiment was already under way. In line with a 1969 AEC decision, the laboratory was renamed at this juncture to honour Enrico Fermi. Although physicists still debate the wisdom of the risks Wilson took with the design of the Fermilab accelerator, even critics agree that because of the ultimate success of the machine, Wilson's design style has set the new standard that major accelerators are built for maximum capability at minimal cost.[77]

4 Conclusions

Cost and resulting pressures clearly forced changes in the way a major accelerator project was planned, organized and constructed. Although Berkeley researchers attempted to plan siting and organize management of the project according to the accelerator building tradition of the 1950s, their attempts were thwarted by both the expense of the project and the altered economic and social environment in the 1960s. Midwestern physicists and politicians pushed for a Midwestern accelerator after the defeat of MURA; young, active experimentalists demanded that the new facility accommodate outside users; and leaders of the physics community sought ways to bolster support for the machine at a time when the federal budget was tightening and the physics community was divided. With the help of AEC Chairman Glenn Seaborg, NAS President Frederick Seitz formed a new management organization, later called URA, which had national representation, an extension of the wide-ranging representation of AUI. To determine the site for the accelerator, a new procedure was devised: the AEC organized a site competition. The new management organization and the site competition promised to soothe the concerns of Midwestern physicists and outside-user advocates and bolster support for the machine. However, in the process, Berkeley designers lost the traditional prerogative of accelerator designers to manage and build an accelerator at the site of their choice.

Continuing economic pressure also worked to defeat the Berkeley team's conservative approach to accelerator design. Impressed by his enthusiasm for creating optimum capability for minimal cost, the URA chose, and the AEC accepted, Robert Wilson as the laboratory's first Director, after LRL physicist Edward Lofgren refused to continue with the project after the Weston site was chosen. Since the URA preserved the tradition of a strong director when defining Wilson's position, Wilson played a dominant role in defining how the new laboratory would be established, including decisions on the design and construction of the machine. Thus, Wilson was able to radically revise the conservative design developed by the Berkeley team, employing a design style which he, ironically, traces to Lawrence's style in the 1930s. After successfully organizing the laboratory to attack problems arising from his much more risky design, Wilson built an expandable machine within the available budget, setting the new standard that major accelerators would be built for maximum capability at minimum cost.

With the advent of site competition, the URA and a new standard of accelerator construction, the new laboratory set a new course for planning, organizing and constructing future accelerator laboratories. The course of the development of accelerator laboratories had changed, because new problems arose as accelerator projects grew larger and more expensive. In the 1950s the continued development of high-energy accelerators hinged on the solution of technical problems — most notably, the development of synchrotrons and strong focusing. In contrast, after 1960, continued development of high-energy accelerators hinged on solving non-technical problems — most notably, bolstering support for the project and adjusting to budgetary constraints. Competing, élite groups of accelerator specialists at Berkeley and Brookhaven facilitated the solution of crucial technical problems in the 1950s. The hegemony of this élite was abolished in the 1960s, when a centrally located accelerator, managed by URA, and built with maximum capability at minimum cost, facilitated the solution of crucial economic and political problems.

In 1963 Leon Lederman imagined a 'Truly National Laboratory', a national resource created to serve the needs of the entire community of high-energy physicists. By 1972 such a laboratory had been established. Although no other laboratory had such a sharp focus on outside users, the new laboratory revised, rather than abolished, tradition. Plans for the new laboratory arose from the momentum of accelerator building generated after World War II in the United States. In addition, the URA was modelled on AUI, and Wilson's design style reflected the influence of Lawrence in the 1930s. Moreover, like its predecessors, Fermilab had a strong Director with the power to shape the research programme.

Just as the formation of the new laboratory was shaped by tradition and new pressures, the funding of the $250 million laboratory forged a new course in the relationship between accelerator builders and the federal government. For the first time, an accelerator project attracted wide-scale attention outside the physics community. Although many physicists were uncomfortable with outside attention and influence, they adapted to the change. It was now necessary to endure the spectacle of the site competition, the push for 'the indoor proton speed record'. The $250 million laboratory was a major national investment, and the pressure for accountability was greater than ever before. The federal government was also forced to adapt as Fermilab was born. For the first time since World War II, the budget was shrinking and increases in science funding prompted public criticism. Despite public scepticism about the value of science and technology and the tightening budget, the federal government funded the $250 million project needed for further exploration of the fundamental nature of matter.

What explains the continuing federal commitment to accelerator building? The momentum for accelerator building generated in the fifteen years after World War II is one factor. No burst of activity was needed to initiate a large new accelerator project. By 1960 physicists expected continued accelerator funding: the AEC was a rich and beneficent patron, and the machinery for assessing and approving accelerator requests was well established.

But the momentum for accelerator building would have certainly died in the 1960s and 1970s without further impetus. During this period, particle physics discoveries continued, and accelerator technology was clearly advanced enough to support further accelerator development. In the 1950s the AEC programme bound the physics community and the federal government into a tight contract cemented by mutual need: the physics community needed money for accelerators, while the federal government needed resources for defence and technology, and the use of accelerators as a symbol of technological, and therefore political, strength.

Did the physics community and government still need each other in the 1960s and 1970s? Owing to the cost of an accelerator in the 200 GeV range, physicists clearly needed the federal government more than ever before. Why the federal government needed physicists is less clear, although the willingness to continue the partnership implies that government leaders were still persuaded by the promise of particle physics. When promoting Fermilab, physicists did not stress the promise of military or technological spin-offs, concentrating instead on the cultural value of exploring the fundamental nature of matter. Since no dramatic spin-offs from accelerator development emerged during this era, it seems unlikely that government leaders made a $250 million investment strictly to advance technology.

Considering the social and economic pressures of the 1960s and 1970s, the government decision to continue its investment in accelerators makes much more sense if it hinged on the desire to bolster the image of the United States as technologically and militarily strong. Further research is needed to assess the power of the accelerator as a national symbol. One way to probe the symbolic value of accelerators would be to examine the dialogue between government representatives and those promoting high-energy physics projects. Does the language used reflect such symbolism? Is the dialogue different from that exchanged when decisions are made about other research expenditures? Additional information could be gained by comparing funding for accelerators and space vehicles — prime symbols of technological strength. Do the two devices share similar funding patterns? Are there similarities in funding justifications? An examination of the founding of Fermilab answers some questions, and prompts many more.

Acknowledgements

It is my pleasure to thank E. Lofgren, G. Seaborg, F. Cole, E. Goldwasser, G. Tape, N. Ramsey, L. Lederman and B. Wilson for documents and editorial advice. Thanks also goes to M. West, R. Meade, S. Weart, J. Warnow and R. Rider for abundant help in archival research. Special thanks to L. Hoddeson, who cheerfully provided information, guidance and encouragement throughout the course of this research.

Notes and References

1. For an account of the history of Fermilab focusing on political considerations see Jachim (1975). A comparison of the history of Fermilab and the Japanese Accelerator Laboratory KEK can be found in Hoddeson (1983). Participant

accounts include Livingston (1968a) and N. F. Ramsey: 'History of the Fermilab Accelerator and URA', Fermilab History Collection, Batavia, Ill., USA. The subject is also covered in Lowi and Ginsberg (1976).

2. For an account of a post-World War I attempt to establish federal support for science see Auerbach (1965) and Tobey (1971). Text references: Kevles (1978), p. 341.

3. Livingston (1969), p. 39, and Schiff (1946), p. 6.

4. Seidel (1983), p. 382; Heilbron et al. (1981), p. 47; Seidel (forthcoming); and Needell (1983), pp. 96–7. Heilbron et al. give an excellent account of the early history of Lawrence Radiation Laboratory. Needell gives an excellent account of the early history of the Brookhaven National Laboratory. Additional information and personal accounts can be gained from Ramsey (1966) and Brookhaven National Laboratory Associated Universities, Inc., The Founding of the Brookhaven National Laboratory by Associated Universities, Inc., Brookhaven, 1948.

5. Kevles (1978), pp. 344–5, 358–9, and England (1982), p. 5. England gives a more detailed description of the formation of the National Science Foundation. For a more detailed account of the formation of the Atomic Energy Commission, see Hewlett and Anderson (1962).

6. Quote as quoted in Pais (1986), p. 19.

7. Kevles (1978), p. 363, and interview with L. Brown, 28 August 1986, Fermilab History Collection, Batavia, Ill., USA.

8. D. Allison: private communication, 27 August 1985.

9. Kevles (1978), pp. 369, 386; and Report by a Special Panel of the President's Science Advisory Committee and the General Advisory Committee to the Atomic Energy Commission: 'Piore Panel Report — 1958'. Appendix 3 in Joint Committee on Atomic Energy, Congress of the United States: High Energy Physics Program: Report on National Policy and Background Information, Washington D.C., 1965, p. 141.

10. Livingston (1969), p. 39; McMillan (1984), p. 31; and E. McMillan: 'The Synchrotron', enclosure to McMillan to E. Lawrence: 4 July 1945, files of G. K. Green, Brookhaven National Laboratory, Upton, Long Island, USA.

11. McMillan (1984), pp. 31, 34; and Livingston (1969), pp. 55, 56, 110. Livingston gives a good overview of the development of synchronous accelerators; for more detailed information, see Livingston and Blewett (1962).

12. A good historical account of the discovery of the muon is Galison (1983), p. 282. Text references: Heilbron et al. (1981), p. 53; Rochester and Butler (1947); Rochester (forthcoming); and Seriff et al. (1950).

13. The limitations faced by cosmic-ray researchers were the following. Although nuclear emulsion provides a dense medium that optimizes the chance of detecting a rare event, the particles produced in associated production have long lifetimes and thus appeared at a distance too far apart to be detected by the nuclear emulsions of the time. The cloud chambers of the time were large enough to detect both particles, but they provided a low-density medium and standard models did not operate continuously. These characteristics created a severe problem for cosmic-ray researchers, who had to rely on the naturally sparse production of strange particles in cosmic-ray showers. Interview with L. Brown, 28 August 1986, Fermilab History Collection, Batavia, Ill., USA. Text references: Livingston (1969), p. 110; Heilbron et al. (1981), pp. 76–81; and Franklin (1979), p. 209. Franklin provides a history of the discovery of parity non-conservation in weak interactions.

14. For a history of the MURA controversy emphasizing political events, see Greenberg (1967). For further insight into the controversy and how it was related to the relationship between the Atomic Energy Commission, Argonne National

Laboratory and various Midwestern universities, see Greenbaum (1971), pp. 243–65. Text quote from F. Mills: private communication, 2 July 1987. Also see Livingston (1969), p. 73.

15. As Brown, Dresden and Hoddeson explain, although MURA scientists were the first to demonstrate a way to accomplish colliding beams, subsequently scientists at Indiana and (independently) at Princeton developed the concept of storage rings for achieving colliding beams. In fact, the storage-ring approach has been the method employed for achieving colliding beams for both electrons and protons. (Brown *et al.* (forthcoming), Introduction.) For more information on MURA technical accomplishments, see Kerst (forthcoming). Text references: Lawrence Jones: private communication, 9 July 1987; and interviews with F. Mills, 13 July 1984, and with D. Judd, 1 May 1984, Fermilab History Collection, Batavia, Ill., USA.

16. After their visit, the CERN delegation, which consisted of O. Dahl, F. Goward and R. Wideroe, brought the news of strong focusing back to Switzerland and plans were quickly made to make the first CERN accelerator strong-focusing. The impact of strong focusing on CERN is discussed in Hermann *et al.* (1987), pp. 213, 273–82. Text references: Livingston (1969), pp. 60–2; and L. Jones: private communication, 9 July 1987.

17. High-energy machines are those which accelerate particles to particularly high energies; high-intensity machines are those which accelerate a large number of particles per second. For more detailed accounts of the design and construction of the AGS, see Blewett (forthcoming) and Courant (forthcoming). Also see Baggett (1980). Text reference: Livingston (1969), pp. 70–3. I wish to thank L. Jones for his lucid summary of the FFAG: L. Jones: private communication, 9 July 1987.

18. Brown *et al.* (forthcoming), Introduction.

19. Interviews with D. Judd, 1 May 1984, E. Lofgren, 3 May 1984, and W. Wenzel, 2 May 1984, Fermilab History Collection, Batavia, Ill., USA.

20. Interviews with E. Lofgren, 3 May 1984, L. Lederman, 20 July 1984, and W. Wenzel, 2 May 1984, and L. Hoddeson interview with F. Seitz, 7 February 1980, Fermilab History Collection, Batavia, Ill., USA.

21. E. Lofgren to E. McMillan, 6 April 1961, files of E. Lofgren, Lawrence Berkeley Laboratory, Berkeley, Cal., USA.

22. California Institute of Technology: 'A proposal to the Atomic Energy Commission for the support of the accelerator design-study program of the Western Accelerator Group', April 1961, pp. 5, 24, Fermilab History Collection, Batavia, Ill., USA; Lawrence Radiation Laboratory: 'Extract from LRL FY1963 budget submission, submitted 4-21-61', files of E. Lofgren, Lawrence Berkeley Laboratory, Berkeley, Cal., USA; and Yuan and Blewett (1961).

23. Sands gives credit to M. Oliphant and T. A. Welton for conceiving cascade schemes. Hoddeson points out that F. Heyn and L. Teng had also made similar suggestions (Hoddeson, 1983, p. 14). Text references: M. Sands: 'Ultra high energy synchrotrons', in Midwestern Universities Research Association: *1959 MURA Summer Study*, p. 1, Library, Lawrence Berkeley Laboratory, Berkeley, Cal., USA; and Sands (1960).

24. H. Gordon, 'Minutes of meeting held to discuss the organization of the study of a super-energy accelerator', and E. Lofgren, 'Notes on a meeting to discuss the organization of a study of a super high energy accelerator', 2 January 1962, files of E. Lofgren, Lawrence Berkeley Laboratory, Berkeley, Cal., USA; and Brookhaven National Laboratory: 'Considerations relating to location of the National Study Group at the Brookhaven National Laboratory', 21 December 1961, and Atomic Energy Commission: 'Initiation of design studies for a proton-synchrotron of several hundred BeV', 22 August 1962, Appendix A, p. 15,

Secretariat Collection, United States Department of Energy Archives, German-town, Md, USA.

25. Interview with A. Tollestrup and R. Walker, 4 May 1985, Fermilab History Collection, Batavia, Ill., USA; and E. Lofgren, 'Conference of McMillan, Lof-gren, Laslett, Kolstad at AEC Germantown, Sept. 26, 1962', files of E. Lofgren, Lawrence Berkeley Laboratory, Berkeley, Cal., USA.

26. Others on the Panel were: P. H. Abelson, O. Chamberlain, M. Gell-Mann, E. Goldwasser, T. D. Lee, W. K. H. Panofsky, E. M. Purcell, F. Seitz and J. H. Williams. Ex officio members were R. M. Robertson from the NSF, who represented the Technical Committee on High Energy Physics of the Federal Council for Science and Technology, and D. Z. Robinson, from the Office of Science and Technology. General Advisory Committee and President's Science Advisory Committee: 'Report of the Panel on High-Energy Accelerator Physics of the General Advisory Committee to the Atomic Energy Commission and the President's Science Advisory Committee'. In U.S. Congress, Joint Committee on Atomic Energy: *High Energy Physics Program: Report on National Policy and Background Information*, Washington, 1965, pp. 85, 104, 108.

27. Quotes from *ibid.*, pp. 94, 89, 90, 95.

28. Quotes, respectively, from *ibid.*, pp. 103–4; and L. Hoddeson, interview with N. Ramsey, 22 January 1980, Fermilab History Collection, Batavia, Ill., USA.

29. At this time billion electronvolts was commonly abbreviated BeV. Since the project is widely known as the '200 BeV', this abbreviation is used here when referring to it. All other energy units are expressed in the more modern term, GeV. Text reference: P. McDaniel to M. Goldhaber, 2 April 1963, files of G. K. Green, Brookhaven National Laboratory, Upton, Long Island, USA.

30. M. Benedict to G. Seaborg, 24 July 1963, and K. Gordon to L. Johnson, files of G. Seaborg, Lawrence Berkeley Laboratory, Berkeley, Cal., USA.

31. MURA's defeat and its effect on the political background for the 200 BeV has been noted by many writers, including Greenberg (1967); Jachim (1975); and Lowi and Ginsberg (1976). All three writers, however, ignore the major contribu-tion of outside-user tensions in setting this background. For further background on congressional interest in science and technology funding and congressional hearings, see Jachim (1975), pp. 82–99; Reagan (1969); Fleming (1965); and Kevles (1978), pp. 413–14.

32. Quotes, respectively, from L. Lederman: 'The Truly National Laboratory', in Brookhaven National Laboratory; *1963 Super-High-Energy Summer Study*, pp. 9–10, Fermilab History Collection, Batavia, Ill., USA. Also see Hoddeson (1983), p. 45.

33. Quotes from E. McMillan to M. White, 20 April 1964, files of E. Lofgren, Lawrence Berkeley Laboratory, Berkeley, Cal., USA. Also see W. Fry to E. McMillan, 26 May 1964, files of E. McMillan, Lawrence Berkeley Laboratory Archives, Berkeley, Cal., USA.

34. Physics Survey Committee, National Research Council: *Physics in Perspective*. National Academy of Sciences, 1972, pp. 645, 646, 648.

35. During this period an AEC budget typically went through several steps. After BoB input the budget was sent to the President. Once approved, the AEC budget was incorporated into the presidential budget, which was presented to Congress. The JCAE then held hearings so that the AEC staff and experts could present their views on proposed AEC expenditures. After incorporating their revisions, the JCAE members took the AEC budget to separate appropriation hearings in the Senate and House of Representatives. G. Tape: private communication, 4 November 1987. For a general description of the budget cycle for federal agencies, see Physics Survey Committee, National Research Council, *Physics in Perspec-tive, op. cit.* [Note 34], p. 655. Text quotes from U.S. Congress, Joint Committee

on Atomic Energy: *Atomic Energy Commission Authorization, FY 1965.* 88th Congress, Second Session, Washington D.C., 1964, p. 1487.

36. Weinberg's article was first published in Weinberg (1963). A detailed explanation of his assessment criteria and other musings on large-scale research can be found in Weinberg (1967). Text quote as quoted in U.S. Congress, Joint Committee on Atomic Energy: *AEC Authorization, FY 1965, op. cit.* [Note 35], p. 1500.

37. Evidence of complaint and the reaction of leaders of the physics community can be found in 'Appendix 17', and 'Appendix 18' in U.S. Congress, Subcommittee on Research, Development, and Radiation of the Joint Committee on Atomic Energy: *Hearings.* 89th Congress, First Session, Washington D.C., 1965, pp. 744 and 751.

38. Quotes from unsigned draft to E. McMillan, 7 October 1964, files of E. McMillan, Lawrence Berkeley Laboratory Archives, Berkeley, Cal., USA. Also see '200 BeV Laboratory', 11 November 1964, files of E. Lofgren, Lawrence Berkeley Laboratory, Berkeley, Cal., USA.

39. F. Cole, C. Dols, F. Gruetter, E. Hartwig, D. Keefe, Q. Kerns, W. Lamb, G. Lambertson, L. J. Laslett, J. Peterson, W. Salsig, L. Smith and G. Trilling to E. Lofgren, 26 October 1964, files of E. McMillan, Lawrence Berkeley Laboratory Archives, Berkeley, Cal., USA; and interviews with D. Keefe, G. Lambertson and L. J. Laslett, 22 December 1986; E. McMillan, 16 May 1984; E. Lofgren, 3 May 1984; L. Lederman, 20 July 1984; and W. Wenzel, 2 May 1984: Fermilab History Collection, Batavia, Ill., USA.

40. Quotes from E. McMillan to P. McDaniel, 23 November 1964, files of E. McMillan, Lawrence Berkeley Laboratory Archives, Berkeley, Cal., USA. Also see W. B. Fowler: 'Meeting at National Academy of Sciences, January 17, 1965, Summary of Notes Taken by Theodore P. Wright', 13 April 1965. L. Hoddeson interview with F. Seitz, 2 February 1980, and interviews with G. Tape, 21 November 1986 and K. Pitzer, 17 May 1984, Fermilab History Collection, Batavia, Ill., USA; and Frederick Seitz: 'National Academy of Sciences Meeting of University Presidents, January 17, 1965'. In U.S. Congress, Subcommittee on Research, Development, and Radiation of the Joint Committee on Atomic Energy, *Hearings, op. cit.* [Note 37], pp. 8–9.

41. Committee members were R. Bacher from Caltech; H. Brooks from Harvard; V. Fitch from Princeton; W. Fretter from the University of California, Berkeley; W. Fry, University of Wisconsin, Madison; J. Gardner, Carnegie Corporation; E. Goldwasser, University of Illinois; G. K. Green, Brookhaven; C. Greenewalt, du Pont; H. Longenecker from Tulane University; E. Piore from IBM; and K. Reed from the NAS. Gardner later withdrew from the Committee when he was appointed Secretary of the Department of Health, Education and Welfare. Site Evaluation Committee: 'The Report of the National Academy of Sciences' Site Evaluation Committee', March 1966, Fermilab History Collection, Batavia, Ill., USA. Text references: G. Seaborg to F. Seitz, 2 March 1965, F. Seitz to G. Seaborg, 6 April 1965, Joseph L. Smith, contract with Frederick Seitz, contract no. AT(49–8)–2783, United States Department of Energy Archives, Germantown, Md, USA; and Atomic Energy Commission: 'AEC–NAS Enter Agreement on Evaluating Sites for A Proposed New National Accelerator Laboratory', 28 April 1965, press release, files of G. Seaborg, Lawrence Berkeley Laboratory, Berkeley, Cal., USA.

42. Atomic Energy Commission: 'Wide Distribution Shown in AEC List of Proposals for 200 BeV Accelerator', 9 July 1965, press release, files of G. Seaborg, Lawrence Berkeley Laboratory, Berkeley, Cal., USA; interviews with Francis Cole, 13 July 1984, John Blewett, 17 November 1986, and James T. Ramey, 21 November 1986, Fermilab History Collection, Batavia, Ill., USA; and P. McDaniel to M. Goldhaber, 2 May 1965, and T. Wright letter to P. McDaniel,

10 June 1965, files of G. K. Green, Brookhaven National Laboratory, Upton, Long Island, USA.

43. G. Seaborg: Diary, 28 July 1965, G. Seaborg to L. Johnson, 28 August 1965, G. Seaborg letter to F. Seitz, 24 August 1965, G. Seaborg: Diary, 1 September 1965. G. Seaborg, record of conversation, 8 September 1965, files of G. Seaborg, Lawrence Berkeley Laboratory, Berkeley, Cal., USA; and H. Busby to Marvin Watson, 8 September 1965, White House Central Files, Lyndon Baines Johnson Library Archives, Austin, Texas.

44. Quote from E. Goldwasser to E. Piore, 13 October 1965, files of G. K. Green, Brookhaven National Laboratory, Upton, Long Island, USA. Also see J. Califano to L. Johnson, 10 September 1965, White House Central Files, Lyndon Baines Johnson Library Archives, Austin, Texas; Atomic Energy Commission: 'AEC Asks NAS to Evaluate 85 Site Proposals for 200-BeV Accelerator', 15 September 1965, press release, files of G. Seaborg, Lawrence Berkeley Laboratory, Berkeley, Cal., USA; and interview with G. Seaborg, 16 December 1984, Fermilab History Collection, Batavia, Ill., USA.

45. Quotes, respectively, from Site Evaluation Committee, 'The Report of the National Academy of Sciences' Site Evaluation Committee', March 1966, pp. 7, 23, Fermilab History Collection, Batavia, Ill., USA. Also see this source pp. 23–39 and Appendix A, p. 11.

46. G. Seaborg, record of conversation, 27 August 1965, files of G. Seaborg, Lawrence Berkeley Laboratory, Berkeley, Cal., USA; J. Califano to L. Johnson, 22 October 1965, White House Central Files, Lyndon Baines Johnson Library Archives, Austin, Texas; and Kearns (1976), p. 296.

47. Quotes from R. Wilson to E. McMillan, 27 September 1965, Fermilab History Collection, Batavia, Ill., USA. Also see Lawrence Radiation Laboratory: *200 BeV Accelerator Design Study, Volumes I and II*. Berkeley, 1965; R. Wilson, 'Some Proton Synchrotrons, 100–1000 GeV', 22 September 1965; L. Hoddeson interview with R. Wilson, 8, 10 May 1978, and 'Report to Board of Trustees by the Scientific Subcommittee', 25 January 1966; and Atomic Energy Commission: 'Review of Program Plans for Higher Energy and Higher Intensity in Accelerator Facilities', 11 February 1966, p. 2, Fermilab History Collection, Batavia, Ill., USA.

48. G. Seaborg, record of conversation, 13 July 1966, and Glenn Seaborg: Diary, 15 September 1966, files of G. Seaborg, Lawrence Berkeley Laboratory, Berkeley, Cal., USA; and H. Traynor to G. Seaborg, J. Ramey, G. Tape, 29 July 1966, Secretariat, United States Department of Energy Archives, Germantown, Md, USA.

49. The view that the decision was made by Johnson as part of a political deal emerged in almost all interviews with physicists, with the exception of Gerald Tape, the one physicist on the Commission. Text references: G. Seaborg, record of conversation, 20 December 1966; Atomic Energy Commission: 'AEC Selects Site for 200-BeV Accelerator', 16 December 1966, press release; C. Mitchell to G. Seaborg, 19 December 1966; G. Seaborg: Diary, 19 January 1967, files of G. Seaborg, Lawrence Berkeley Laboratory, Berkeley, Cal., USA; 'Weston's flawed windfall', *New York Times*, 20 December 1967; D. Judd to G. Seaborg, 10 February 1967, files of E. McMillan, Lawrence Berkeley Laboratory Archives, Berkeley, Cal., USA; J. Erlewine to G. Seaborg, W. Johnson, S. Nabrit, J. Ramey and G. Tape, 21 November 1966, Seaborg Collection, United States Department of Energy Archives, Germantown, Md, USA. (I wish to thank Robert Seidel for providing the 10 February 1967 document.)

50. On 21 June 1965 articles of incorporation were filed for the Universities Research Association, Inc. See L. Bacon: 'Minutes of First Meeting of Board of Trustees of Universities Research Association, Inc.', 16 September 1965, files of E. Lofgren,

Lawrence Berkeley Laboratory, Berkeley, Cal., USA. Text references: Universities Research Association: 'Proposal for Continuing Studies for a 200 BeV Accelerator Facility', 23 December 1966, and L. Hoddeson interview with N. Ramsey, 26, 27 February 1980, Fermilab History Collection, Batavia, Ill., USA.

51. Quote from E. Lofgren to N. Ramsey, 12 January 1967, files of E. Lofgren, Lawrence Berkeley Laboratory, Berkeley, Cal., USA. Also see interview with E. Goldwasser, 10 July 1985; L. Hoddeson interview with N. Ramsey, 26, 27 February 1980; N. Ramsey: 'Universities Research Association, Inc.: History and Issues', 17 January 1973; Universities Research Association: 'Board of Trustees, Minutes of Meeting, January 15, 1967', Fermilab History Collection, Batavia, Ill., USA.

52. D. Judd to G. Seaborg, 10 February 1967, files of E. McMillan, Lawrence Berkeley Laboratory Archives, Berkeley, Cal., USA, and G. Seaborg: Diary, 11 February 1967, files of G. Seaborg, Lawrence Berkeley Laboratory, Berkeley, Cal., USA.

53. For an example of promotional efforts for high-energy physics, see Yuan (1965).

54. Seaborg, the first scientist to head the AEC, saw promotion of basic research as an important part of his mission as AEC chairman. For an early example of Seaborg's view on basic research, see President's Science Advisory Committee: *Scientific Progress, the Universities, and the Federal Government*, Washington D.C., 1960.

55. U.S. Congress, Joint Committee on Atomic Energy, United States Congress: *Hearings*. 90th Congress, First Session, Washington D.C., 1967; E. Lofgren: private communication, 16 February 1988; Garren *et al.* (1967), which was submitted on 14 April 1967. (The author wishes to thank Lloyd Smith for providing the 1967 document.)

56. Quotes from 'Vs. scientific luxury', *New York Times*, 16 July 1967, editorial. Also see U.S. Congress, Joint Committee on Atomic Energy: *Atomic Energy Commission Authorizing Appropriations, FY 1968*. 90th Congress, First Session, Washington D.C., 1967; Norton *et al.* (1982), p. 953; and L. Johnson to G. Seaborg, 26 July 1967, United States Department of Energy Archives, Germantown, Md, USA.

57. N. Ramsey to R. Wilson, 6 February 1967; and interviews with E. Goldwasser, 15 May 1987, and R. Wilson, 1 April 1987, Fermilab History Collection, Batavia, Ill., USA.

58. Interviews with D. Keefe, G. Lambertson, L. J. Laslett, 22 December 1986, and John Blewett, 17 November 1986; D. Getz, untitled manuscript, May, 1977; R. Wilson: 'Some Aspects of the 200 GeV Accelerator', 12 September 1967; Lillian Hoddeson interview with Robert Wilson, 12 January 1979, Fermilab History Collection, Batavia, Ill., USA.

59. Interview with R. Wilson, 25 May 1987, Fermilab History Collection, Batavia, Ill., USA.

60. Quote from interview with R. Wilson, 25 May 1987, Fermilab History Collection, Batavia, Ill., USA.

61. Sanford (1976); Don Getz, untitled manuscript, *op. cit.* [Note 58]; R. Wilson, 'Some Aspects of the 200 GeV Accelerator', *op. cit.* [Note 58]; Cole (1971), pp. 1, 2; interview with Francis Cole, 29 August 1986; Fermilab History Collection, Batavia, Ill., USA.

62. Don Getz, untitled manuscript, *op. cit.* [Note 58]; 'Questions Raised on the Design of the 200 BeV Accelerator', files of Frederick Mills, Fermilab, Batavia, Ill.; and interview with Francis Cole, 29 August 1986, Fermilab History Collection, Batavia, Ill., USA.

63. Quote from L. Hoddeson interview with N. Ramsey, 26, 27 February 1970, Fermilab History Collection, Batavia, Ill., USA. Also see Atomic Energy

Commission: 'Atomic Energy Commission Summary Notes of 200 BeV Accelerator Briefing', 1 September 1967, files of G. Seaborg, Lawrence Berkeley Laboratory, Berkeley, Cal., USA; L. Lederman to W. Salsig, 8 November 1967, files of W. Salsig, Lawrence Berkeley Laboratory Archives; Livingston (1968b), p. 27; N. Ramsey: 'History of the Fermilab Accelerator and URA', 2 May 1975, and National Accelerator Laboratory: 'Construction Project Data Sheet, Schedule 44', Draft, 20 September 1967, Fermilab History Collection, Batavia, Ill., USA.

64. National Accelerator Laboratory: *Design Report*. Batavia, Ill., USA, 1968, pp. 3-2.

65. As Cole (1971), p. 9, explained, the new septum 'uses thin wires as the inner electrode of a transverse electric field. This electrode can be made much thinner than the current-carrying septum of a magnetic deflector and, consequently, the particle losses on it are much smaller.' The septum was crucial to single emergent beam extraction because higher particles losses would have produced unacceptable radiation levels. Text references: National Accelerator Laboratory: *Design Report, op. cit.* [Note 64], pp. 3-3 to 3-10; Lawrence Radiation Laboratory: *200 BeV Accelerator Design Study, Volume I, op. cit.* [Note 47], p. I-1; and interview with F. Cole, 29 August 1986, Fermilab History Collection, Batavia, Ill., USA.

66. The LRL report promised a machine producing 30 trillion protons per pulse, while the NAL report promised a machine producing 50 trillion protons per pulse. The reduced-scope machine mentioned in the 13 December 1965 report was estimated at $257 million. National Accelerator Laboratory: *Design Report, op. cit.* [Note 64], p. I-1; Lawrence Radiation Laboratory: *200 BeV Accelerator Design Study, Volume I, op. cit* [Note 47], p. I-1; and E. Lofgren, 'On the Costs of an Accelerator With Reduced Initial Capabilities', 13 December 1965, files of E. Lofgren, Lawrence Berkeley Laboratory, Berkeley, Cal., USA. Text references: National Accelerator Laboratory: *Design Report, op. cit.* [Note 64], pp. 3-10, 3-13, 3-14; Lawrence Radiation Laboratory; *200 BeV Accelerator Design Study, Volume I, op. cit.* [Note 47], p. I-7.

67. Quotes, respectively, from L. Hoddeson interview with N. Ramsey, 26, 27 February 1980, and interview with F. Cole, 29 August 1986, Fermilab History Collection, Batavia, Ill., USA. Also see Lawrence Radiation Laboratory: *200 BeV Accelerator Design Study, Volume I, op. cit.* [Note 47], p. III-10; Lawrence Radiation Laboratory: *200 BeV Accelerator Design Study, Volume II, op. cit.* [Note 47], p. VI-11; and National Accelerator Laboratory: *Design Report, op. cit.* [Note 64], p. A-4, Figure 8-1.

68. NAL designers report many instances of Wilson's penchant for spartan standards. As J. R. Orr remembers, Wilson was very pleased when he visited the construction of an experimental area and noted that one piece of equipment had been fashioned out of lead bricks piled on a discarded bookshelf. Many stories are also told of Wilson's attempts to spur progress by promoting competition among his staff. For example, P. Livdahl remembers that Wilson used linac development, which was rapid, since Wilson employed a design similar to one already developed at the Los Alamos Meson Facility and Brookhaven, as a 'whip' to goad other groups into fast progress. Quotes, respectively, from J. R. Orr: private communication, 29 January 1988; and Hoddeson interview with Philip Livdahl, 15 May 1978, Fermilab History Collection, Batavia, Ill., USA. Text quote from interview with Edwin Goldwasser, 15 May 1987, Fermilab History Collection, Batavia, Ill., USA.

69. U.S. Congress, Subcommittee on Research and Development and Radiation of the Joint Committee on Atomic Energy: *Hearings.* 92nd Congress, Second Session, Washington D.C., 1972, pp. 1433.

70. National Accelerator Laboratory: 'The Village Crier', 1970, 2, Fermilab History Collection, Batavia, Ill., USA.

71. Wilson now feels that the lack of air-conditioning had no effect on the magnet

problem. R. Wilson: private communication, 1 February 1988. Text quotes, respectively, from Universities Research Association: 'URA Annual Report 1971, January 1972', in U.S. Congress, Joint Committee on Atomic Energy: *Hearings, op. cit.* [Note 69], pp. 2053 and 1435. Also see this source, p. 1436, and interview with F. Cole, 29 August 1986, F. Cole: 'Monthly Report of Activities', 31 October 1971, pp. 1-2, F. Cole: 'Monthly Report of Activities', 31 January 1972, p. 2, Fermilab History Collection, Batavia, Ill., USA.

72. Quotes, respectively, from Universities Research Association: 'URA Annual Report 1971', *op. cit.* [Note 71], p. 2053; and interview with R. Wilson, 1 April 1987, Fermilab History Collection, Batavia, Ill., USA. Also see F. Cole: 'Monthly Report of Activities', 31 October 1971, p. 3, Fermilab History Collection, Batavia, Ill., USA.

73. Quotes, respectively, from Universities Research Association: 'URA Annual Report 1971, January 1972', *op. cit.* [Note 69] and interview with H. Edwards, 13 March 1987, Fermilab History Collection, Batavia, Ill., USA. Also see F. Cole: private communication, 7 January 1988, Fermilab History Collection, Batavia, Ill., USA.

74. Quote from interview with H. Edwards, 13 March 1987, Fermilab History Collection, Batavia, Ill., USA. Also see interviews with J. R. Orr, 12 March 1987, R. Wilson, 1 April 1987, and E. Goldwasser, 15 May 1987, Fermilab History Collection, Batavia, Ill., USA.

75. As a joke, F. Cole reported that the 200 GeV beam was achieved on 30 February, 'to make it look better on paper'. In line with this, the monthly report for February was also dated 30 February. F. Cole: private communication, 11 January 1988. Text quote: F. Cole: 'Monthly Report of Activities', 31 January 1972, p. 1, Fermilab History Collection, Batavia, Ill., USA. Also see R. Wilson to N. Ramsey, 29 October 1971, R. Wilson: 'Formation of the Accelerator Section', memorandum to the staff, 21 October 1971, F. Cole: 'Monthly Report of Activities', 31 December 1971, pp. 1–3, and F. Cole: 'Monthly Report of Activities', 30 February 1972, p. 1, Fermilab History Collection, Batavia, Ill., USA.

76. Quote from interview with J. R. Orr, 12 March 1987, Fermilab History Collection, Batavia, Ill., USA. Also see interview with E. Malamud, 12 March 1987, and F. Cole: 'Monthly Report of Activities', 31 March 1972, p. 1, Fermilab History Collection, Batavia, Ill., USA; and U.S. Congress, Joint Committee on Atomic Energy: *Hearings, op. cit.* [Note 69], p. 1435.

77. L. Hoddeson reveals that Wilson used the $20 million to support research on building a superconducting accelerator. Hoddeson (1987), p. 35. (This source provides an excellent history of the Fermilab 'Energy Doubler.') Text references: Atomic Energy Commission: 'AEC Names 200 BeV Accelerator in Honor of Enrico Fermi', 29 April 1969, press release, files of G. Seaborg, Lawrence Berkeley Laboratory, Berkeley, Cal., USA; R. Nixon to G. Seaborg, 18 April 1969, Seaborg Files, United States Department of Energy Archives, Germantown, Md, USA; and interviews with D. Keefe, G. Lambertson and L. J. Laslett, 22 December 1986, and F. Cole, 29 August 1986, Fermilab History Collection, Batavia, Ill., USA.

12

The CERN Beam-transport Programme in the Early 1960s

John Krige

The proton synchrotron at CERN — the European Organization for Nuclear Research — first reached its design energy of 25 GeV on 24 November 1959. This achievement was the culmination of seven years of research and development effort directed to endowing Europe with the biggest particle accelerator in the world. It heralded the transition to a new phase in the laboratory's life — the transition from a period in which the machine was *constructed* to one in which it was to be *exploited* as a useful tool for high-energy physics. An accelerator is only a means to an end. To do physics with it, it has to be surrounded with auxiliary equipment — equipment to generate secondary beams, to 'transport' them away from the main accelerator, to study their reaction products, and to analyse the resulting data. This auxiliary equipment, said E. Lofgren (Berkeley) in 1956 'is extremely important and is needed on a scale commensurate with the accelerator'. Indeed, at the time the CERN proton synchroton (PS) was commissioned, the laboratory administration was suggesting that it would cost about as much over the next four years to set up and run the PS experimental programme (about 96 MSF — million Swiss francs) as it had cost to build the machine itself (about 100 MSF).[1]

It is something of a commonplace among the international high-energy physics community that CERN 'failed' to exploit the PS properly in the early 1960s, that its preparation of the experimental programme was 'poor', lacking both scientific focus and 'adequate' experimental equipment. This was openly said inside the organization itself at the time, where it reflected the deep disappointment and concern felt by CERN physicists at being 'beaten' to major new discoveries by American rivals, particularly the Brookhaven National Laboratory (BNL), which had commissioned a new machine (the AGS) similar to the CERN PS in July 1960.[2] It has been confirmed by Irvine and Martin, who have used citation analysis to illustrate 'the full effects of [CERN's] not being ready to mount a comprehensive experimental programme. . .'. They show that between 1961 and 1964 publications from the Brookhaven AGS were cited about four times more frequently than those produced at the CERN PS, and that Brookhaven scientists published about three times as many highly cited papers as did those in Geneva.[3]

Irvine and Martin summarize a number of reasons given by European and American scientists for why CERN was 'late' and for the difference in scientific performance between it and Brookhaven, in particular. In the eyes of the community CERN could not compete at this time because American physicists were (allegedly) more competent, more experienced, more bold and imaginative, better managers, etc., than those in Europe. In this chapter I want to go further than these spontaneous and rather general 'analyses' made by the participants themselves. More specifically, I want to try and study in a little more depth one aspect of the so-called equipment gap in the first PS experimental programme: the lack of adequate beam-transport material. To this end, we first describe quickly what a beam-transport system is. To refine the problem further I then look at CERN's first order for beam equipment, comparing it with the corresponding decisions at Brookhaven. With a better idea of the nature of the beam 'crisis' in Geneva in 1961/early 1962, I conclude with a discussion of its causes. The major finding is that the 'equipment gap' arose because CERN scientists placed far too low a priority on the aquisition of beam-transport equipment, so leaving the laboratory singularly poorly placed to exploit the potential offered by having the most powerful accelerator in the world ready before that of its main American rival.

1 What is a Beam-transport System?

Most of the nuclear physics experiments around a high-energy accelerator are performed, not with the primary accelerated beam, but with secondary beams of quite different particles. The simplest way of creating such particles was to intercept the primary beam with a target — usually Be or Al foil or wire in the case of the CERN PS — which was flipped up into its path. Apart from the elastically scattered protons, the main particles emitted from the PS and the AGS in this way were positive and negative pions (i.e. π^+, π^-) and kaons (i.e. κ^+, κ^-), and antiprotons. As reported by Cocconi (CERN) in Rochester in 1960, the kaons constituted 10–30 per cent of the pions produced at the target, and the antiprotons some 1 per cent of all particles emitted up to about 10 GeV. All these particles were produced in useful quantities over a wide range of momenta and of angles to the position of the target.[4] The aim of a beam-transport system is to select the desired particles from this mix.

Until the mid-1950s this was achieved by suitably combining two basic elements: quadrupole magnets or 'lenses' to focus the beam and bending or analysing magnets to select particles of a particular momentum from it. The 'output' of a beam-transport system was thus a mix of particles with the same momentum. This did not matter very much if the particle of interest was the most abundant component, and/or one's detectors could discriminate between particles on the basis of velocity, as could electronic detectors (e.g. counters). With the advent of bubble chambers in the latter half of the 1950s, neither of these conditions was satisfied: the interest turned to studying interactions of the far less abundant kaons and antiprotons with hydrogen in the chamber, and the cleaner the incident beam the easier would be the subsequent data analysis.[5] As a result, essentially two ways of producing purified beams were developed. In the simpler, the so-called electrostatic separator, a momentum-analysed beam passes at right angles through a uniform electric field established between two parallel plates. The resulting separation of the particles is proportional to the field strength and length of the electrodes, and inversely proportional to the momentum of the particles. In the technically more complex radiofrequency separators, which were still very much in the design stage in the early 1960s, separation was achieved by having 'unwanted particles travel with the velocity of a suitable wave in the structure [so being] deflected out of the way, whereas the wanted ones slip in phase by 2π in their passage and so receive no net angular deflection'.[6]

Now that we have some idea of what is involved in building a purified secondary beam, let us look at a schematic of a relatively simple specimen: the separated 1.5 GeV kaon beam built at the CERN PS, which was first used with a bubble chamber in December 1961 (see Figure 12.1).[7] The beam contained one 10 m-long electrostatic separator; eight quadrupole lenses, Q_1–Q_8; and three analysing magnets, M_1–M_3. The overall length from the target to the bubble chamber was 41 m. Every 100 κ^- at the detector were 'contaminated' with about 8 π^- and 50 μ^-, the last produced by the decay-in-flight of the negative pions.

2 The First Order for Beam-transport Equipment, 1958–9

Around September 1958, after more than a year of discussing the equipment needed to exploit the new accelerator properly, the CERN senior staff agreed that 'the experimental work on the CERN-PS [could] best be carried out by experimental teams led by competent physicists, each with its own budget and a staff suitable for constructing the apparatus and carrying out the experiments'. Five kinds of teams were envisaged, each associated with a particular kind of detector. Apart from being responsible for their detectors, each team was expected to 'work out the requirements of its experiments' in terms of targets, beam-transport systems, shielding, and so on, although 'the design and ordering of the actual apparatus could probably best be done by the PS engineers'. Co-ordination between the various teams was one of the tasks of a so-called Executive Committee for the PS Experimental Programme, which first met on 23 September 1958.[8]

By the end of November two worked-out proposals for beams had been

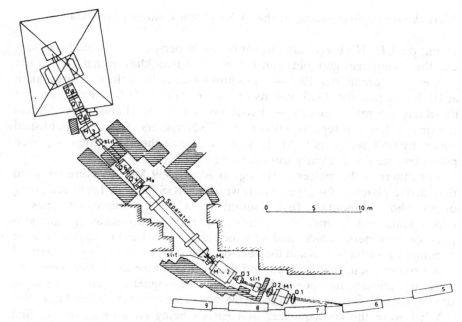

Figure 12.1 Layout of K_1 Beam in the North Hall of the CERN PS.

placed before the Committee. Guy von Dardel, who was doing electronic experiments on the CERN 600 MeV synchrocyclotron, proposed setting up an unseparated antiproton beam to be used primarily in conjunction with Cherenkov counters and time-of-flight techniques for velocity selection. The other proposal was made by Colin Ramm's team. Ramm was an engineer in John Adams's PS group, and had been given the responsibility of building CERN's 1 m propane bubble chamber. Deciding not to restrict themselves to the needs of their detector, Ramm's group laid out five beams which filled the entire south experimental area, where the first experiments were to be done. At the heart of their system lay a separated antiproton beam for bubble-chamber work, whose position was chosen so as to have the greatest path length (i.e. antiproton momentum) in the south hall. The purification of the beam was to be achieved in three successive electrostatic separator tanks, and Ramm's group estimated that it would provide antiprotons up to 2.5 GeV/c. The separators were to be built in-house.[9]

The Executive Committee agreed to order the magnets needed to build Ramm's and von Dardel's beams. After further discussion inside the PS group of the design of the elements and the generators needed to power them, the details of an order for 44 magnets (18 bending and 26 quadrupole) were placed before the Finance Committee on 25 May 1959. Having surveyed the market and called for tenders, the CERN staff proposed that the order be placed with the Swiss firm Oerlikon.[10]

3 Brookhaven's First Beams for the AGS — Some Comparative Data

To put the CERN's early planning for beams in perspective — and because it was the 'equipment gap' between Geneva and Brookhaven that caused the greatest consternation in 1961 — let us now look briefly at the measures taken at BNL to equip the AGS with its first set of beams. Our sources here are inevitably poorer in quality — Brookhaven National Laboratory Annual Reports, a few trip reports written by CERN visitors, and some published papers by BNL scientists.[11] All the same, whatever their shortcomings, three points emerge rather clearly from these various items.

First, there is the matter of timing: in April 1959 Mervyn Hine reported that 'active planning for experiments with the AGS has only [sic] been going on for about 6 months', that Courant and Cool had 'discussed types of experiments and beams with interested parties in Brookhaven, and also possible customers outside, and with the results have tried to decide on a set of generally useful beams and the necessary bending and focusing magnets'.[12] It is striking, then, that BNL started seriously planning its AGS beams, and were in a position to place their first order for magnets, at about the same time as CERN — even though the AGS was some 8 months behind the PS.

What were the consequences of CERN's being (relatively) late in first thinking about the beams it needed to launch its PS experimental programme? There are two points to be made here. First, it probably contributed to the physicists making a poor survey of the beams available from the new machine in the first 6 months of 1960. Cocconi was to complain strongly about this limitation, and with some justification: Brookhaven apparently ensured that it had enough bending magnets (which was the only type needed to do the survey) when they were needed.[13] The second point to make, and one that is more important for our purposes here, is that the timing of CERN's order for beam-transport equipment did *not* apparently delay the start of the PS experimental programme as such in any substantial way. After the PS first 'worked', an enormous amount of time had to be spent on 'tuning' the machine and surveying the beams, with the result that, up to August 1960, so for the first 9 months of the PS's life, there was only a total of about 500 hours of machine time available for experiments. By the time the programme got under way in a serious manner (late summer 1960), the bulk of the first order of magnets had been delivered.[14] Developments at BNL tend to confirm this conclusion. For while people there may have *planned* their beams relatively earlier than their CERN colleagues, they did not apparently *order* the equipment until later. Indeed, by February 1961 — so 8 months after the AGS first 'worked' and some time after CERN's experimental programme had started — the BNL staff were still surveying their beams, and had taken delivery of all their bending magnets, but only some of their quadrupoles, and were testing one of their separators.[15]

The second striking difference between Brookhaven's and CERN's first plans is the stress which the former placed on building a flexible separated beam. Whereas the centrepiece of Ramm's layout was a clean antiproton beam of up to 2.5 GeV/c — and a very simple, very low-energy (500 MeV/c) unseparated kaon beam was tucked away in one corner of the south hall —, Cool and Courant proposed building a beam which could supply separated antiprotons up to 5 GeV/c *as well as* separated positive and negative kaons of a reasonably high energy ('in principle' up to 3.5 GeV/c; a more realistic limit waited on the beam survey). To achieve this performance, Cool and Courant proposed using two stages of separation (only one was foreseen in Ramm's antiproton beam), and pointed out that the beam was 'arranged such that for use at lower momenta various separators could be removed to shorten the total length of the system, which would be desirable for use with lower momentum κ mesons'.[16] Brookhaven also ordered a much greater variety of bending and quadrupole magnets. CERN restricted its first order to two standard lengths (1 m and 2 m), and one aperture for each type (about 25 cm width for the bending magnets, 20 cm diameter for the quadrupoles). In September 1961 a CERN visitor to BNL reported that the laboratory had now taken delivery of three types of bending magnets and four types of quadrupoles of varying length and aperture. The details are given in Table 12.1, where the reader should note, in particular, that BNL had far *fewer* magnets on hand at the AGS than had CERN at the time, and that it had ordered several lenses which were shorter than 1 m. We shall need this information later.

My third point: what about the separators? In 1959 both laboratories designed long antiproton beams which involved coupling three 10 m-long electrostatic separators in series. However, there were two important differences. I have already mentioned one of them: that it was foreseen to use two such triplets in the BNL beam, and only one in the CERN beam, i.e.

Table 12.1 The first order of magnets placed by CERN and BNL, being a comparison of the equipment available in each laboratory around September 1961 for use with their strong-focusing machines.

	Bending magnets	Quadrupole lenses	Total
CERN	5 of 25 cm × 1 m 13 of 25 cm × 2 m	18 of 20 cm × 1 m 8 of 20 cm × 2 m	18BMs, 26QLs
BNL	4 of 15 × 45 cm × 90 cm 5 of 15 × 45 cm × 180 cm 3 of 15 × 75 cm × 180 cm	5 of 20 cm × 120 cm 4 of 30 cm × 75 cm 6 of 30 cm × 150 cm 6 of 15 × 60 cm × 90 cm	12BMs, 21QLs

NOTES
1. We have assumed 1 in = 2.5 cm.
2. Data for CERN from CERN/FC/357, 15/5/59; data for BNL from F. Grütter's report (see Note 11). The fourth type of lens at BNL was a rectangular quadrupole designed by Panofsky; in his report (see Note 11) Schnell gave their length as 90 cm and specified that they were for use with beam separators.

Brookhaven planned to acquire six separators to CERN's three. Second, whereas Ramm's group went ahead with building these 10 m-long devices at CERN, Brookhaven changed their plans and decided to order six shorter separators (15 ft, so about half as long) of the 'Cosmotron type modified to operate with hot glass cathodes'.[17] The BNL Annual Report dated 1 July 1961 claimed that 'the first prototype unit [of this device] has been delivered and is being assembled in preparation for testing'. Three months later Baltay *et al.* presented a paper in which they described BNL's 3.3 GeV/c antiproton beam and indicated that it had just begun to be used for experiments; two of these 'conventional' separators were included in the layout.[18] At this stage, by contrast (i.e. early September 1961), Ramm's device was not yet operating. Indeed the first of his 10 m-long electrostatic separators first ran in a similar beam 3 months later (i.e. December 1961), after 6 months of 'difficult and painstaking' work.[19]

4 The Beam-transport 'Crisis' at CERN in 1961

Now that we have an idea of the different equipment available at CERN and at Brookhaven in 1961 we are better able to refine the nature of the problems faced by the Genevan physicists that year — problems which surfaced explicitly during the meeting of the Scientific Policy Committee at the end of April. The most voluble of their number on that occasion were the bubble-chamber scientists, who were particularly frustrated at not being able to launch a good programme in strange-particle physics — 'the main cause' of which, said Research Director Gilberto Bernardini, was the 'unsatisfactory beam situation'.[20]

Strange particles, to quote Pickering, was the name given to 'unstable, strongly-interacting particles, or "resonances"'.[21] A few had been discovered in the early 1950s; with the commissioning of the 6 GeV Berkeley Bevatron in 1954, their numbers increased considerably, and by the early 1960s the population explosion of resonances had, to quote Pickering again, 'transformed much of the HEP landscape'. Table 12.2 gives one some idea of these particles and the kind of equipment that was needed to produce and to detect them. Strange-particle physics evidently called for the widespread use of hydrogen bubble chambers in separated beams of antiprotons and kaons.

A lack of beam equipment stifled strange-particle physics at CERN in 1961 in two specific ways. First, the laboratory was *short of separators sufficiently powerful to work at the new energy range* it had opened up. Ramm's 10 m tanks were not available and CERN could only work on resonances at all because they were able to borrow two 3.5 m-long electrostatic elements offered to them by Cresti of Padua University in September 1960.[22] With this equipment they could build a separated antiproton beam with a design momentum of 0.6–1.5 GeV/c. This was frustratingly low: it did not exceed the separated antiproton momenta achievable with the Bevatron (1.61 GeV/c —

see Table 12.2, item ω^0) and was much below the 3.3 GeV/c antiprotons that were available at Brookhaven from September onwards. Furthermore, with the Cresti separators tied up in the antiproton beam, and without Ramm's separators, CERN could not build up a separated kaon beam — and when it did (early 1961), the maximum attainable energy was again disappointingly low (1.5 GeV). This was no higher than what the Bevatron could achieve (K$^-$ of 1.8 GeV/c — see Table 12.2, item Ξ^*), considerably below what Brookhaven had available during the first half of 1962 (2.5 GeV/c — see Table 12.2, item Ξ^*, second entry), and below what CERN Track Chamber Committee Chairman Gregory called the kaon momentum 'specific of our machine' (i.e. 1.5–2 GeV/c).[23] In short, because CERN had not enough adequately powerful separators at its disposal in 1961, it was not able to start a strange-physics programme which exploited the potential of its new accelerator, and which enabled it to compete with the work of this kind being done from the autumn onwards at Brookhaven.

The second major difficulty CERN faced, particularly at the end of 1961/early 1962, was the *shortage of magnets*, both in kind and in number, to

Table 12.2 Beams and detectors used to discover some of the most important resonances with accelerators at the end of the 1950s and early 1960s[a]

Date[b]	Particle	Laboratory	Detector[c]	Particle	Beam momentum (GeV/c)	Number of separators
9/2/59	Ξ^0	LRL	15 in HBC	κ^-	1.15	2
31/10/60	Y^*	LRL	15 in HBC	κ^-	1.15	2
16/2/61	κ^*	LRL	15 in HBC	κ^-	1.15	2
12/5/61	σ meson	BNL[d]	20 in HBC	π^+	0.9, 1.09, 1.26	1
11/5/61	σ meson	BNL[d]	14 in HBC	π^-	1.89	?
12/5/61	Y_0^*	LRL	15 in HBC	κ^-	1.15	2
31/5/61	Y_0^*	LRL	15 in HBC	κ^-	1.15	2
14/8/61	ω^0 meson	LRL	72 in HBC	$\bar{p}$	1.61	3
15/9/61	ω^0 meson	LRL	72 in HBC	$\bar{p}$	1.61	3
10/11/61	η^0 meson	LRL	72 in HBC	π^+	1.23	?
2/1/62	η^0 meson	LRL	15 in HBC	κ^-	0.76, 0.85	2(?)
19/2/62	Ξ	CERN	81 cm HBC	$\bar{p}$	3.0	1
19/2/62	Ξ	BNL	20 in HBC	$\bar{p}$	3.3	2
15/6/62	2 neutrinos	BNL	Spark ch.	π^+, ν	—	—
27/6/62	Ξ^*	LRL	72 in HBC	κ^-	1.8	2
2/7/62	Ξ^*	BNL	20 in HBC	κ^-	2.24, 2.5	2
22/8/62	f^0 meson	BNL	20 in HBC	π^-	3.0	2
2/11/62	f^0 meson	CERN	1 m HLBC	π^-	6.1	0(?)

[a] This list has been compiled from the chronology given by Six and Artru (1982), which contains '*all* resonances found in the years 1960–1961, but only some typical ones in the following years'.
[b] The date is the date on which the paper in which the finding was published was received by the journal — see references given in Six and Artru.
c HBC: hydrogen bubble chamber;
HLBC: heavy liquid bubble chamber.
[d] These two results were obtained using the external proton beam of the Cosmotron.

build good beams *with the separators it had*. This may sound surprising when we remember that CERN seems to have had *more* magnets at this time than did Brookhaven — but then it must not be forgotten that, whereas BNL had *one* two-stage separated beam producing *both* antiprotons and kaons, CERN could not 'convert' the one into the other, and had to build different beams for each. In April 1961 this led Gregory to fear that, once Ramm's separators were available for building an antiproton beam, so releasing Cresti's for a kaon beam, the two would 'involve using half the bending magnets and quadrupoles available at CERN for bubble chamber work and greatly reduce the number of magnets and quadrupoles used by counter groups to date'.[24] The problem was compounded by CERN's not having enough *flexibility* in the choice of elements available to build beams: as Hine put it in 1961, 'the layout and efficiency of some beams would be improved if we had more lenses less than 1 m long and also separators intermediate between the 3 m tanks and the new 10 m tanks'.[25] This was particularly noticeable, for example, when the kaon beams were built in the north hall early in 1962. The separators available apparently limited them to low momenta (1.5 GeV/c maximum), which meant that the beam had to be short if all the kaons were not to decay in flight before they reached the detector: the beam shown in Figure 12.1 was about 40 m long, while a typical antiproton beam would be closer to 100 m in length. Short beams called for short lenses — and to build them CERN made copious use of eight quadrupoles borrowed from Saclay, which were considerably shorter than the Geneva laboratory's standard 1 m lenses.[26]

In 1961, then, CERN bubble-chamber physicists found it difficult if not impossible to embark on a competitive physics programme, primarily because they lacked sufficiently powerful separators to enter the energy region opened up by their 28 GeV machine, and because the lack of variety in their 'standard' lenses imposed severe limitations on the quality of the beams they did build. The situation was summed up by Hine in May 1961 in the gloomiest possible terms: 'no magnets or separators, no money, no staff and no clear ideas of what is needed'.[27] How had it arisen?

5 The Roots of the CERN Beam-transport 'Crisis' in 1961/Early 1962

I believe that the single most important reason for the crisis was that no one at CERN was willing or able to give sufficient priority to the design and construction of beams, or to dedicate themselves to reacting to developments in a rapidly changing field, to foreseeing where modifications were called for and, in general, to ensuring that the PS would be equipped with good secondary beams when it worked. The task was marginal to the main professional concerns of both engineers and physicists at the time, and devolved on the shoulders of people who devoted a considerable amount of time and effort — too much, one might say — to some components (separators), while neglecting others (magnets).

The engineers, by and large, did not regard designing secondary beams as their responsibility. Their main task was to get an intense, high-energy *primary* beam circulating in the machine; in so far as secondary beams concerned them at all, they saw their main role as developing the means to produce them inside the machine, and to extract and transport them up to the shielding wall surrounding the main ring: once the beam entered the experimental hall, it was no longer their business. They could help design and perhaps even build the individual *components* of a beam (magnets and separators); the *kind* of beam to be built, its design and layout, what components were needed and how they were to be combined, depended on the physics one wanted to do, and was primarily the physicists' job.[28]

The physicists, for their part, were not particularly interested either (with the exception of von Dardel, as we have seen). They wanted to do physics with a new accelerator which was proving to be far more exciting than originally thought, not concern themselves unduly with beams for a machine that was 2 years off working. And when they did put their minds to the task, they tended to think in terms of rather simple unseparated beams of the kind and on the scale they had grown accustomed to in working with counters and small bubble chambers around the low-energy machine.

What of the group that first laid out most of the beams for the south hall, drew up the specifications for and ordered the first set of magnets, and designed and built the first electrostatic separators — the group, subsequently division, led by Ramm? Their *main* task, their top priority, was to build a 1 m propane bubble chamber to be used around the PS; prevailing policy dictated that they *also* had to design the beams that were to feed it.[29] Professionally oriented towards building heavy, technically challenging equipment — they were essentially engineers who had been recruited by Adams to build the PS; intellectually isolated — no one had yet been found to take on the job of building CERN's 2 m hydrogen bubble chamber and to think of the beams it needed; and pressured by deadlines — the PS was due to work in about a year and delivery times for beam transport magnets ranged from 12 to 18 months; they had not the inclination, the experience or the time to devote themselves seriously to designing beams. This was of secondary importance, a responsibility *deriving* from their propane chamber work, something that had to be got out of the way. So, for example, at the international conference held at CERN in September 1959 we find that not one of the nine papers in a session devoted to the 'Production, Transport and Separation of Particles' was given by a member of CERN — and that Cocconi was the only Genevan who even participated in the formal discussions of papers by Blewett, Courant and Cocl, Livingston, Panofsky, Ticho from Berkeley, Brookhaven, Stanford.[30] And with the details of the order settled, we find one suggestion after the other — for more magnets (Cocconi), for a few spare C-shaped magnets (Yuan), for very-high-field magnets (Lofgren), for special 30 cm Panofsky-type lenses (Hine) — firmly rebuffed.[31] In short, for Ramm and his team the design of secondary beams and the decision on the elements needed for them was something that had to be done, essentially a distraction from more pressing and interesting tasks, a duty rather than part of an evolving intellectually and technically challenging project.

In fact the only part of the beam-transport project that fired the imagination of Ramm's engineers was the electrostatic separators. Conceived on a big scale, the possibilities provided by building 10 m tanks were far richer than those offered in the design of standard bending and analysing magnets. And they were exploited to the full. The vacuum chamber in the separators in the Brookhaven 3.3 GeV/c antiproton beam which was used to discover the antihyperon in December 1961/January 1962 (see Table 12.2, item Ξ) was a little under 1 m in diameter and 5 m in length. The gap between the separating plates was 5 cm and the maximum operating potential across them (i.e. before sparking) was about 400 kV.[32] The separator used in the parallel experiment at CERN — one of three similar devices built by Ramm's group — was far bigger and more complicated. Its vacuum tank was 1.4 m in diameter and 10 m long, the gap between the electrodes was variable between 6 cm and 26 cm, and it could be 'made to fit a diverging or converging beam, or a curved mean trajectory in the case of low energy particles'. Provision was made to place 1 MeV across the electrodes at large spacings.[33] Whereas Brookhaven built separators of a conventional type based on a design in use at the Cosmotron — which were ready about a year after the machine worked —, CERN engineers rose to the challenge of building a 'perfect', 'all-purpose' device — which was ready about 2 years after the machine worked. The lack of good beams at CERN in 1961 arose because they 'chose' not to build electrostatic separators which were simpler, less flexible, and quicker and easier to design, construct and commission; indeed it was only because Cresti's tanks were available that CERN was able to build any separated beams in 1961 at all.

Now that I have presented and defended my main hypothesis as to why there was a beam transport 'crisis' at CERN in the early 1960s, I want to consider two rival explanations of the phenomenon. The first is that given by Francis Perrin, a leading member of the French scientific establishment, member of the CERN Council, and delegate to the Scientific Policy Committee. Speaking during the debate in the SPC in April 1961, he remarked that 'the shortage of beam transport equipment was clearly due to the budget limitations imposed by the Member States'.[34] Now it is true that at this time a number of Member States — notably the British — were disturbed by the rising trend in CERN's budgets. Having originally been led to believe that the annual costs of the laboratory would actually fall off after the construction period was over, they were now finding that it was going to be just as expensive, if not more so, to exploit the new machines. In an attempt to control the situation, the Council decided in the late 1950s to impose ceilings on the laboratory's expenditure — 195 MSF being the figure for 1960 to 1962 inclusive.[35] And while this policy undoubtedly caused some 'hardship' at CERN, particularly in 1961, the financial argument cannot, in my view, do the work that Perrin asks of it. For one thing, there is very little evidence to suggest that anyone placing orders for beam equipment — or at least for standard equipment — felt particularly constrained by financial concerns, and this is so whether we are speaking of the first order, placed in 1959, or the second, placed in 1960. For another, CERN actually spent quite a lot of money on beam-transport equipment and shielding for the PS between 1959 and 1962 — about 35 MSF —, which was almost twice as much as they had originally budgeted for in 1959, and about the same as was spent in these 4 years on track chambers and on equipment for electronic experiments (about 30 MSF each).[36]

The second alternative explanation we want to consider is that put forward by Mervyn Hine at the SPC meeting in 1961, and repeated in a recent interview with me: that the lack of 'small separator tanks' and 'small magnets' had arisen because 'the beam transport orders that were placed . . . were based on an assumption that it was to be mainly high-energy beams, they were designed . . . with the idea of doing high-energy physics rather than low-energy physics'.[37] This idea not only derives its plausibility from *the fact* that the beams being built at the time were low-energy beams and were designed around the use of small separators and short magnets (both borrowed). It also resonates with a number of other widely held views about why CERN physicists were scientifically 'unprepared' when the PS started working, e.g. that, generally speaking, physicists find it difficult to predict what they will be doing and what equipment they will need to do it two or three years ahead, that the mental horizons of CERN physicists were set by the scale of the 600 MeV synchrocyclotron so that they could not imagine what working at 28 GeV entailed, and that for want of 'experience' they were unable to draw up a proper physics programme to exploit the PS and could do no more than simply repeat and refine what their American colleagues were doing.

Before tackling this question, we must first clarify what is to be understood by a high-energy beam. In the context of what was scientifically urgent and technically possible at the time, it meant having separated beams of anti-protons of about 3 GeV/c and of kaons of about 2.5 GeV/c — so not particularly high, even by the standards of the mid-1960s, but, crucially, usefully above the beam energies available at the 6 GeV Bevatron. That granted, the first point to note is that Ramm's group did *not* design high-energy beams of this type: as we have said, they envisaged a separated antiproton beam of up to 2.5 GeV/c and an unseparated kaon beam of a few MeV. Second, it is not true that physicists were initially uninterested in having high-energy beams. Even as Ramm was planning his beams, emulsion physicists complained that the energy of the kaon beam was too low — and were apparently ignored. At the end of 1960 a survey of proposed track chamber experiments included requests for antiprotons of $\approx$2–3 GeV/c and for negative kaons $\geq$1.5 GeV/c. In April 1961, as we have seen, Gregory regretted that the available beam energies were below the threshold 'specific' to the CERN PS. And in October 1961 the representatives of eight British universities 'hoped that both high-energy and low-energy p beams would be produced [at CERN] in 1961', going on to say that, in addition to working with low and intermediate energy ('up to about 1.5 GeV') kaons, it was 'well worthwhile attempting to produce a separated K^- beam at 2–3 GeV or even higher energies'.[38] In short, physicists — or at least track chamber physicists — worked with low-energy beams at CERN in 1961, because *they had no other choice, given the separating power available on site at the time.* The beams were not built to satisfy their experimental proposals; the proposals were adjusted to the (disappointing) reality of the equipment that was there. And while they found interesting, if not particularly exciting or novel, things to do at antiproton and kaon energies no higher than those at Berkeley, they looked forward to the day when it would be possible to build good, separated 'high-energy' beams. They first had them in 1962.

One last point by way of conclusion. The two hypotheses I have just considered, while differing considerably in richness — Hine's is obviously more stimulating than Perrin's 'classical' appeal — have a feature in common which they share with the 'analyses' for CERN's relative 'backwardness' unearthed by Irvine and Martin in their interviews — they are participant's interpretations of events. And what is striking about both is how tied they are to conjunctural circumstances, to the specific landscape of the context in which they were first uttered, be those the acute budgetary difficulties CERN was facing in 1961 (Perrin) or the array of low-energy beams in use in the south hall (Hine). Starting from what seemed evident, and situating it in a framework of beliefs and impressions, each 'spontaneously' built his own interpretation of where the roots of the beam-transport problem lay. What both lacked, however — what almost any participant inevitably lacks — was the ability, the possibility, of detaching themselves from the present, of seeing how and why it had been shaped by the past. That is the historian's privilege, and I hope to have turned it to good account in this chapter.

Acknowledgement

I wish to thank Mervyn Hine for a useful discussion on this chapter.

Notes and References

All documents cited below are available in the CERN archives. When a source is given in brackets (e.g. DIRADM. . .) it refers to a particular collection of papers (Director of Administration). Official documents in the CERN/. . . series and internal reports are usually classified by their series number in the archives.

1. Lofgren (1956). For the estimate of the cost of the CERN PS (100 MSF including buildings and general expenses) see Kowarski (1959), p. 383. The estimated cost

of the PS experimental programme (96 MSF over the years 1959–62) is given in Finance Committee document CERN/FC/327, 24/1/59.

2. See particularly the meeting of the Scientific Policy Committee held on 29 April 1961, minutes CERN/SPC/133/Draft, 7/7/61, and Director-General Weisskopf's report to the 21st Council Session, in CERN Council Minutes, 13/6/62, pp. 14 *et seq*.

3. Irvine and Martin (1984).

4. Cocconi (1960), p. 802.

5. Crozon (1987), Chap. 6, and Galison (1987) give good descriptions of bubble chambers and electronic detectors and the different practices they demand.

6. M. G. N. Hine, *Report on a Visit to the U.S., April 1959*, Report PS/Int. TH 59-7, 12/5/59, p. 42.

7. Amato *et al.* (1963).

8. See CERN/SPC/71, 1/9/58. Dominque Pestre has given a detailed description of the evolution of CERN's organizational structure in his analysis of the 'CERN system' (forthcoming).

9. For von Dardel's and Ramm's proposals see CERN/Ex.C/13, 20/11/58, and CERN/Ex.C/14, 21/11/58, respectively (DIRADM 20417).

10. John Adams presented a list of magnets to the Executive Committee at its meeting on 10 February 1959 — minutes are CERN/Ex.C/33, 18/2/59 (DIRADM 20417). The recommendation to the Finance Committee is CERN/FC/357, 15/5/59.

11. The sources we have used are the Brookhaven Annual Reports dated 1 July 1960, 1961 and 1962, at pp. xxviii, 23 and 32, respectively, trip reports by CERN staff — M. G. N. Hine in April 1959 (see Note 6), C. A. Ramm, *Notes on Visit to Brookhaven on 6th, 7th September 1960*, Report PS/Int.EA 60-9, 12/9/60 (KHR22138), W. Schnell, *Notes on a Visit to Brookhaven, February 2 to 7, 1961* (KHR22138), and F. Grütter, *Beam Transport Magnets and Power Supplies at Brookhaven, September 1961*, Report ENG/Int.DL 61-15, 13/10/61 (LK22457) — and published articles: Courant and Cool (1959), Baltay *et al.* (1961, 1963) and Leitner *et al.* (1963). One should not be misled by the publication date of these last two papers; they were presented at a conference held at CERN from 16 to 18 July 1962, whose proceedings were published in the journal.

12. Hine's report is cited in Note 6; the quotations are on p. 39.

13. Cocconi made his comments in his (1960); C. A. Ramm reported (see Note 11) that by September 1960 Brookhaven had received 'some bending magnets and rectangular aperture quadrupoles'; Schnell confirmed the availability of bending magnets in February 1961 (see Note 11).

14. Cocconi (1960), p. 799, gives information on the useful time of the PS and on magnet delivery dates; for the latter see also CERN Annual Report, 1960, p. 13.

15. This information is derived from Schnell's trip report (see Note 11).

16. The quotation from Courant and Cool (1959) at p. 407.

17. The quotation is from the paper presented by Baltay *et al.* (1961) at Brookhaven. We do not know precisely why this group, which did not include Courant and Cool (who had designed the first beam layout), used shorter separators than those envisaged in the original plans; several reasons spring to mind (to be sure that the equipment was ready on time, to allow for more flexibility in beam layouts. . .), all of them, of course, derived from what we have established independently and tending to reinforce the image of BNL as supple, well-prepared, etc. Perhaps it was simply due to shortages of staff or money.

18. The quotation from BNL's 1961 Annual Report is at p. 23, where the number of separators is also given, and where they are described as 'conventional'. See Baltay *et al.* (1961, 1963) for a description of this antiproton.

19. CERN Annual Report 1961, p. 103.

20. The published minutes of the SPC meeting of 29 April 1961 are CERN/SPC/

133/Draft, 7/7/61; we have supplemented them by transcribing additional material from a tape recording of the entire proceedings. Bernardini made his comment on the first page of his *Report on the Present Status of Research at CERN*, CERN/SPC/128, 20/4/61, which was laid before the meeting.

21. Pickering (1984). The quotations in this paragraph are all on p. 33.

22. See letter Cresti to Adams, 1/9/60, and the reply, 20/9/60 (DG20573).

23. On the tape of the proceedings of the SPC on 29 April 1961 — see Note 20. The *design* momentum of the antiproton beam is given in Appendix 1 to Bernardini's report quoted in Note 20; in fact CERN apparently used only very low energy antiprotons (below 200 MeV) in hydrogen bubble chamber work in 1961 — see Hine's paper on *Statistics on Research Collaboration Between CERN and Other Countries*, CERN/SPC/213, 28/11/65, pp. 37-8. The energies of the kaon beams are from Amato *et al.* (1963) and Aubert *et al.* (1963).

24. See CERN/SPC/133/Draft, 7/7/61, p. 9.

25. In his *Note on Beams and Beam Transport for the CERN PS 1961-63*, CERN/SPC/ 137, 19/7/61.

26. The Amato beam (see Note 23) used two 25 cm-long lenses, one that was 50 cm long, and five CERN 1 m quadrupoles. The Aubert beam (Note 23) used four quadrupoles 22 cm long, and one that was 1 m long.

27. Letter Hine to Weisskopf, 8/5/61 (DG20576).

28. See Note 8 for information on the perceived role of engineers in beam building.

29. Ramm's determination to build a 1 m propane chamber is described in Krige and Pestre (1986).

30. The proceedings are referred to in Note 2. The session in question was one of seven, which gives one an idea of the importance which at least some members of the community attached to the topic.

31. These various rebuffals are to be found in the minutes of several meetings of the Executive Committee for the PS experimental programme — namely, documents CERN/Ex.C/33, 18/2/59, CERN/Ex.C/42, 17/6/59, CERN/Ex.C/48, 11/8/59, CERN/Ex.C/54, 22/10/59.

32. The information on the BNL separators is extracted from the papers by Baltay *et al.* (1961, 1963).

33. For information on the CERN separators see Ramm (1960), and Germain and Tinguely (1963).

34. CERN/SPC/133/Draft, 7/7/61, p. 9. Perrin's remark followed a somewhat more nuanced intervention by John Adams which was not included in the official minutes.

35. I have studied the procedures for 'controlling' the CERN budget in Chapter 12 of Volume 2 of Hermann *et al.* (forthcoming).

36. These figures are derived from the functional breakdown of expenditure given in CERN Annual Report, 1962, p. 142.

37. The quotation is from the tape of the 19th meeting of the SPC held on 29 April 1961. The official minutes, CERN/SPC/133/Draft, 7/7/61, make no reference to this argument.

38. These requests for 'high-energy' beams are to be found, respectively, in *Emulsion Work with the CERN 25 GeV Proton Synchrotron* (Geneva: CERN 59-13, 1959), pp. 7-8, the draft minutes of a meeting on track chamber experiments at CERN held on 21/12/60, document TC/COM 61/4, 17/1/61, the tape of the 19th meeting of the SPC held on 29 April 1961, and the proposal made by eight British laboratories for bubble chamber experiments at CERN during 1961, document CERN/TC/COM 61/43, 26/10/61.

13

'Monsters' and Colliders in 1961: The First Debate at CERN on Future Accelerators

Dominique Pestre

CERN, the European Organization for Nuclear Research, was officially set up in September 1954.[1] Two years before, it had already been decided by the provisional organization to build a 600 MeV synchrocyclotron (SC) and a multi-GeV proton synchrotron (PS). Both formed the first generation of CERN accelerators. The SC went into operation in 1957; the PS of 28 GeV, at the extreme end of 1959. One year later, a spirited debate about the next generation of machines for CERN started in Europe. The aim of this chapter is to describe how the debate occurred in that first year. Let us note before starting that only scientists took part in the discussion — which is a key reason why I wish to study it carefully here. The debate involving government officials only started in 1962/3.[2]

The chapter is divided into two main parts. The first is a chronological account of what happened — so it is of a monographic nature. The second is of a more reflective kind.

1 A Description of the Debate

The discussion on new accelerators is a continuing process in the field of high-energy physics. The main reason for this is that at the world level there always exist people who have finished building one machine and start thinking seriously about another. The second reason is that physicists always try to improve the existing machines and that it is out of the question not to keep in touch with the latest innovations.

At CERN there was an interest in new machines well before November 1959, when the PS produced its first beams. In fact, the Accelerator Research Group (AR group) was set up at the end of 1956 when CERN people thought that Europe was lagging behind the USA and the USSR. The initial task — and, in fact, that which occupied the AR group up to 1959 — was to investigate novel ideas for future accelerators, to keep in contact with everything new happening in the world. For this the group studied the acceleration of protons by a plasma ring, a concept developed by Russian physicists, the Fixed Field Alternating Gradient (FFAG) type of proton synchrotron advocated by the Midwestern Universities Research Association, and beam stacking in storage rings.[3] At the end of 1959, the AR group was studying experimentally two particular projects as far as accelerators were concerned. One, in decline, dealt with plasma rings; the other, in full expansion, was a complex electron model, a two-way FFAG synchrotron of 100 MeV, able normally to accelerate and stack, in the same machine, two beams going in opposite directions and intersecting inside the machine at several points.[4]

In 1960, the PS of 28 GeV being finished, the AR group's situation changed. From being somewhat marginal it became a division endowed with some of the best engineers who had built the PS.[5] In the course of the year the direction of the work was seriously reconsidered: the studies on plasma were definitely abandoned, as was the electron model. Both were considered too far removed from immediate application on a larger scale, too complex to be used in the short term. Instead, a simpler electron model allowing one to study beam storing in a ring was agreed to. In parallel another group prepared a full-scale project of two intersecting storage rings (ISRs) for the protons emitted by the 28 GeV PS.[6] Technically, this means that one injects alternatively the particles in two independent rings of magnets which intersect with one another in one or more places; one stores the particles in each of them up to the maximum number possible — which should be at least sufficient to produce a rate of interaction in the intersecting regions which is high enough for experimental purposes. The advantage of this technique is that the energy of the collision is very high (the energy released when two 21.6 GeV protons collide frontally in an intersecting zone is 'equivalent' to that released by a 1000 GeV proton hitting a fixed target). One clear disadvantage is the low rate of interaction due to the far lower density of a beam compared with that of a solid or liquid target.

What were the reasons for this change of orientation in the AR division? The first was the prospect of a collaborative venture between the USA and the Soviet Union to build in common a giant accelerator — what we could call the 'competitive aspect'; the second was the reinforcement of the AR division by the enterprising engineers who came from the PS division — i.e. an organizational feature. These two reasons help us to understand why there was a push for a project that could be put into practice rather quickly — and why less conventional and not immediately applicable schemes were abandoned. To understand why *proton ISRs* were thought of — rather than something else — we have to realize that storing was considered as an elegant and challenging (technically speaking) way of going up in energy, that it belonged to a tradition of research in the CERN AR division, and that the very good but unexpected intensity of the CERN PS would make it a perfect injector for any storage rings. Note that this cluster of reasons combines new political and institutional factors (competition, reorganization), particular traditions of scientific thought, a coincidental element (the way the PS worked) and the preferences of the AR personnel.[7]

On 22 December 1960 the AR division proposed officially that CERN should study the *building* of ISRs for 25 GeV protons. Early in 1961 the debate on this proposal got under way. It remained rather calm up to March and became very animated from April to June — when it quietened down. The debate occurred between two main groups that the actors themselves recognized as such: the 'builders' and the 'physicists' — the first group being essentially the engineers who built the PS; the other, the experimentalists using the accelerators, along with the theoreticians. Nuances have to be mentioned: Colin Ramm of the ex-PS division thought, for example, that the ISRs were too small a project for CERN; on the other hand, some physicists liked the idea of the ISRs, and clear differences of opinions appeared between generations. Nevertheless, the main line of division, the enduring one, was that between what I call 'builders' and 'physicists'.[8]

First question What caused the animated discussion in the spring? It was the fact that in March the Director-General asked the Scientific Policy Committee (which had to advise the Council of Member States' representatives on any decision for investment) to *decide* on the proposal put forward by the AR division in December 1960. Hence the violent reaction from those hostile to the ISRs, or unwilling to decide *now*. For them, the move was unfair and too sudden.[9]

Second question Did this move reveal anything of the power structure at CERN? Yes: it showed the importance and the ability to take initiatives of the ex-PS division led by Adams. In 1960 it represented half of the staff of CERN, it was in charge of all the heavy tasks and it enjoyed a justified renown. Quick action on their part was then always possible. It would be wrong, however, to see what they did simply as a manoeuvre to *impose* their preference on the others: rather they were sure that the ISRs were a very good solution for CERN, a solution which would be agreed to by a majority.

Third question What were the terms of the debate? For the AR people, ISRs would permit high intensities in each of the storage rings at 25 GeV; they would allow very high energy (more than 1000 GeV in p–p interactions) in the intersecting regions; they would be a technical challenge which had to be mastered (the ISRs as the paradigm, the 'only way' for the future). For the physicists, the first important feature of any new accelerator was not the energy in the centre of mass of the incident proton and the target proton but its ability to produce good (i.e. intense and of rather high energy) secondary beams (anti-protons, π, kaons, . . .). Unfortunately, they claimed, ISRs were only a 'p–p experiment' without any possibility of getting very intense and well-collimated secondary beams with which rare events could be studied. Second, there was little or no flexibility in the way of setting up the experiments around the collision zones of the machine. Their preference was therefore for an accelerator using a fixed target and distributing beams at will in experimental halls.[10]

Fourth question   Were there other kinds of arguments than the 'scientific' ones? Certainly, and they were always mixed with the others in the discussion. Indeed, scientific arguments were rarely used in isolation. Costs were mentioned — the ISRs being cheap; availability — the ISRs could be ready before any other machine; location — part of the debate was that the ISRs could only be constructed in Geneva, whereas a big fixed-target synchrotron would have to be built elsewhere; and international context — was there to be a 300–1000 GeV intercontinental synchrotron?[11]

What was the result of this discussion? It was a decision of principle: since the machines were to be built for physicists, they had to speak first. They had to say what they needed, what suited them best 'from the scientific point of view'. From that starting point, the AR people would see what was possible. In other words, *ends* had to be defined first, the *means* to achieve them being studied afterwards. As a result the Directorate withdrew its proposition for a quick decision by the SPC, and a conference on theoretical aspects of very-high-energy phenomena was called for at CERN.[12]

The conference was duly held in June 1961. It confirmed that flexibility (secondary beams) and accuracy in measurement (intensity requirement and ease of setting-up of experiments) were to dominate the choice; and that there was probably not too much to be gained from a physics point of view by an important jump in energy. In any event, increase in energy could not be achieved at the expense of intensity. In agreement with these conclusions, it was agreed that, instead of the ISRs, a 50–100 GeV, very-high-intensity proton synchrotron (i.e. 10^{13-14} protons per second — p/s) should be studied at CERN by the AR division. It was also accepted that a final decision on construction be postponed to the end of 1962. In the meantime, the priority was to use the 28 GeV PS to the maximum and to seek for intercontinental collaboration.[13]

No sooner had these conclusions been reached in July than events across the Atlantic forced people at CERN to think again. During the summer, three important gatherings were organized in the USA. In 1959 some officials of the American and Soviet Atomic Energy Commissions had met and suggested collaboration in high-energy physics. In 1960 they had decided to study an intercontinental accelerator for 300–1000 GeV protons, as we have said. Brookhaven National Laboratory was in charge for the USA. In May 1961 BNL issued a report discussed at length in August. In parallel another seminar was held at Berkeley on 100–300 GeV proton synchrotrons. Finally, in September the third international conference on accelerators was held at Brookhaven. European representatives took part in all three meetings.[14]

The most important result of these various gatherings was that the image European physicists and builders got of what a synchrotron of several hundreds of GeV would look like changed. Some technical problems were overcome or likely solutions suggested; some new ideas were put forward, such as the possibility of straight sections; some delicate questions were systematically tackled — in particular, the way(s) to achieve high intensities (let us say 10^{13} p/s) with these very-high-energy PSs. If no definitive answer was given on what the *best* technical solution would be, various alternatives were judged more than promising by all experts.

Two features struck the Europeans. First, that the overwhelming majority of the American scientists (the division between physicists and builders was not so strong in the USA) were convinced that a very-high-energy/high-intensity PS was the *obvious* multipurpose accelerator of the next generation: its secondary beams, in particular, would be energetic and of very good intensity — of an intensity even better than with a 50 GeV, 10^{14} p/s PS. Second, these same Americans argued that it was crucial to go up in *energy*. Their reasons were partly based on the kind of studies which were undertaken in June 1961 at CERN — namely, which kind of experimental information would be decisive for the theoreticians, etc. — but also on a totally different type of argument, i.e. that you do not need particular reasons to go up in energy! Nature was too complex to be anticipated, they claimed, and it had always been 'generous' with physicists in the past. This seems indicative of an important cultural difference between Europe and the USA in the early 1960s — in particular, in attitude towards a big and costly project. After all, it is no coincidence that up to the end of 1961 the Europeans called machines in the hundreds of GeV range 'monsters'.[15]

The net result of all this was rather simple: one month after the international conference in Brookhaven, many European scientists — and those at CERN, in particular — argued that the first piece of equipment to be studied by the AR division had to be a 300 GeV PS of high intensity (some 10^{13} p/s). Three main reasons appear when reading carefully the documents of the time. First, a 300 GeV PS was a technical challenge — like the ISRs — which was capable of raising enthusiasm. Second, the fact that the Americans were sure to build one such machine in the next few years implied the necessity of an equivalent big piece of equipment in Europe. Third, it appears that many European scientists genuinely recognized that such a PS was one of the best, if not *the* best multipurpose accelerator that could be thought of. These three motivations have to be perceived as forming a whole, as constituting an ensemble the actors did not disentangle — and not as separate entities. For example, if the 300 GeV PS was a perfect machine from the point of view of competing with the USA, it was precisely because it was considered by the European physicists, after the American studies of the summer, as the most complete and powerful tool for the physics for the 1970s.[16]

In November 1961, however, the Directorate of CERN preferred not to narrow its future possibilities of choice. It decided to please the various groups — the 'ISR enthusiasts' had not disappeared — and gave as a programme of work to the AR division the design study of both a 300 GeV PS *and* the ISRs.[17]

2 Reflections on the Debate

Now that we have an idea of what happened in this small slice of time cut from a complex process, I would like to step back from the events and consider two problems. One is of a methodological character — and I deal with it shortly; the other concerns the specificity of high-energy physics in Europe in the early 1960s.

What should be noted first is that CERN appears to have been in a kind of second position compared with the USA during this period — and this is not to be taken as a value judgement. CERN was created in 1952–54 precisely to catch up with America. In this respect, the PS was a tremendous success, but it was not an end in itself and it was not the most complicated part of the programme. Certainly, the PS was a sophisticated machine, but to build it one had to deal only with *one* kind of problem, a technical one, with regard to which the Europeans were quite capable. On questions which required a combination of the talents of numerous groups possessing very different know-hows, on the contrary — such as the discussions about long-term equipment or, more importantly, the choice of an experimental programme — CERN was still not too sure of itself, CERN was still lacking in managerial skills and long-term planning methods, CERN was still very young.

The difficulty, however, was not only that Europe rather lacked experience in dealing with these highly integrated jobs. It also lay in the traditional difference in investment which separated Europe and the USA. In brief, the Americans had spent, were spending and were mentally ready to spend more than the Europeans. The regular growth of the expenditures in the field was considered usual in America — which was not yet the case in Europe. Clearly, in the early 1960s the race was 'imposed' on the Europeans, and without the constant push ahead coming from America, the European physicists would have gone at a quieter pace and would have considered stopping for a while: after all, their work was quite expensive and they knew that that had to be considered.[18]

A third noteworthy element in understanding the specific situation of Europe is to realize that there were in the USA several high-energy physics laboratories able to compete with each other and with CERN. Therefore, the problem never was, for an American laboratory, to decide for *the best* equipment for the country. It was only considering its own equipment, and that in the light of the competition it was engaged in with others. The tendency was, in general, to bet on a particular type of device — MURA and the FFAG machines, BNL and the 300 GeV PS with a linac as injector, . . . —, to envisage the biggest one of that type and to fight for money — those three things being done in the same move. In contrast, CERN had to be the best in Europe, what it decided had to be *the* European solution and it had to cover its bets against *all* the American laboratories. For this reason, everything had to be thought of and thought of again. Hence the far longer scientific discussions, always disconnected from the financial ones with political authorities — at the beginning of such a process Council members often said, in the early 1960s, that raising money was their job and that scientists had only to think 'from a scientific point of view';[19] hence the often more complex devices built at CERN — after all, the more sophisticated your apparatus, the more polyvalent its possible usages (think here of the electrostatic separators mentioned by John Krige). Hence, finally, the tendency to be more cautious, perhaps too conservative in the final choices made.

The second point I would like to stress now is very classical — at least for those who have been trained as historians of general history. It advises one to be careful and to avoid quick interpretations, notably those which have their roots — consciously or not — in strong hypotheses, which tend to favour systematic sets of presuppositions, which reflect the same 'reading', the same 'grille de lecture' imposed on any kind of primary material. Let me draw attention to two such possible 'readings'.

The first one would be the 'rationalistic way' of explaining things — in our case, the way scientists debate, the way a decision-making process operates. The actors who generally appear in these kinds of description are: (1) perfectly informed before saying or deciding anything — what we could call the presupposition of total information; (2) infinitely sensitive to their environment and to the consequences of their actions — what could be called the presupposition of transparency (no Freudian unconscious, for example); (3) fully rational in their moves — i.e. they can define clearly at any time their objectives, they are able to classify their means, and they can compare them with their ends and, of course, make the one best choice at the end of a linear process. Note here that the traditional image of the scientist at work corresponds to this description.[20]

According to this way of perceiving the world, the story just related would have been essentially concerned with pure intellectual debates, with the state of the art in science and technology; it would have recalled the ever-present necessity of increasing the energy of every new accelerator (which is 'obvious' nowadays); it would have envisaged the logics of the arguments — and the coherence of the decision taken in the light thereof. Be sure of one thing: if you know your documentation well enough, and because the debates I have discussed reflect an increasing ability to control nature technologically, you could make a very strong case. You could show, in particular, that the 'rest' of the story — builders versus physicists, for example — was marginal: in the long term, and through the hesitant path of a human history, you got an optimal choice. Among other things, you could demonstrate that it was the technical nature of the debate of summer 1961 in America which led to the final solution; that it was its intellectual content which, in the last instance, was the decisive element.

With the same set of documents, it would also be possible to write a purely social history of that debate. The latter would dismiss the strict scientific content of the arguments as being irrelevant as such to the output, and would look more closely at the concrete relations of power, at the way what is called 'knowledge' or a 'decision' was generated, negotiated and accepted by a majority as constituting 'the one best solution'. It would consider the scientist as a professional with a particular training and set of interests belonging to a particular generation, marked by specific references and scientific preferences. In our case, it would stress the existence of the two main groups, engineers vs academics; the way they organized themselves to win support — this was consciously done; the complex game of power between the physicists, who had a superior symbolic power, notably that of speaking 'from the scientific point of view', and the accelerator people, who were more influential inside the organization. Be sure of something here too: it would not be difficult to show that the argumentation was very dependent on the social position inside the organization and on the moment at which it was used (it is easier for an accelerator expert having built a 30 GeV PS to speak about a 300 GeV machine than it is for a physicist who has still not used such a 30 GeV PS); it would not be difficult to show that the dividing line, if altered slightly, did not really change between 1961 and 1965 — the builders went on defending the ISRs; finally, it would be very easy to show that the decision ultimately reached in 1963/4 — ISRs *and* 300 GeV PS — was typical of a well-known type of process, which Warner R. Schilling has identified as 'how to decide without actually choosing'.[21]

Why have we developed all this? To show that both of these approaches — they are not the only ones — throw real light on the *process* but that it would be fatal — even if reassuring, because of the feeling of coherence — to analyse it from the point of view of only one of them. In fact, and this is now rather banal, the received division between 'intellectual' or 'rational' understandings and 'social' or 'external' explanations has not to be treated as necessarily useful nor given once and for all. As Martin Rudwick has expressed it, the production of technical and scientific knowledges 'can be regarded as both artifactual and natural, as a thoroughly social construction that may nonetheless be a reliable representation of the natural world'.[22]

In more general terms, what we wanted to do was to recall some old historical precepts. First, not to reduce too quickly, not to violate the *messiness* of what human beings do when they 'think' or 'decide'. 'La poussière des détails événementiels ne s'explique pas par des réalités éternelles: gouverner, dominer, le Pouvoir, l'Etat', wrote Paul Veyne. Second, to accept that several logics are at work in human affairs and that they are not necessarily reducible, one to the other. One has then no choice but to immerse oneself in the historical *process*, and not to consider the result as already contained in the premises — a rule which François Furet has illustrated so brilliantly in the case of the French revolution.[23] Finally, to accept the consequences when the documentation resists one's favourite interpretation, and not to add too much intelligibility, too much necessity — be it of a rational or of another kind — to the account.[24]

Notes and References

1. Hermann *et al.* (1987).
2. D. Pestre, *La seconde génération d'accélérateurs pour le CERN, 1956–1965, Etude historique d'un processus de décision de gros équipement en science fondamentale,* Geneva, CERN-CHS 19 report, 1987.
3. Documents CERN/SPC/36, 18/2/57; CERN/SPS/58, 17/2/58; CERN/SPC/63/ Draft, 2/4/58 (CERN archives).
4. Documents CERN/SPC/87, 31/3/59; CERN/SPC/92, 8/5/59 (CERN archives).
5. M. J. Pentz, *Accelerator Research at CERN, 1956–1967,* Geneva, CERN-68-9 report, 1967, pp. 20–1.
6. Barbier *et al.* (1959). H. G. Hereward, K. Johnsen, A. Schoch and C. J. Zilverschoon, *Present Ideas on 25 GeV Proton Storage Rings,* Geneva, CERN PS/Int. AR/60-35 report, 22/12/1960.
7. CERN Council, minutes 1–2/12/59, p. 61; 14/6/60, pp. 27–8; 8–9/12/60, pp. 72–3; Documents CERN/SPC/123, 17/2/61, and CERN/378(B), 10/11/60 (CERN archives).
8. A careful study of this aspect is in Pestre (1988).
9. Documents CERN/SPC/125, 14/4/61; CERN/SPC/125/Add., 14/4/61 (CERN archives).
10. Documents CERN/SPC/125, 14/4/61; CERN/SPC/133/Draft, 7/7/61; correspondence in file DG20576 — in particular, letter Van Hove to Director-General, 13/6/61 (CERN archives).
11. For a good example, see the debate reported in *International Conference on High Energy Accelerators,* US-AEC, 1961, pp. 287–90.
12. Document CERN/SPC/133/Draft, 7/7/61 (CERN archives).
13. *International Conference on Theoretical Aspects of Very High-Energy Phenomena,* held at CERN, 5–9/6/61, Geneva, CERN 61-22, 1961.
14. *The Berkeley High-Energy Physics Study, 15/6–15/8/61,* Berkeley, LRL, UCRL-10022, 1961; *Design Study for a 300–1000 BeV Accelerator,* Upton, BNL report, 28/8/61; *Int. Conf., op. cit.* [Note 11].
15. Confront Serber, an American theoretician, in *Int. Conf., op. cit.* [Note 11], pp. 3–5, and UK *DSIR and NIRNS Joint Consultative Panel for Nuclear Research,* 24/10/61 (CERN archives, DG 20831).
16. Documents CERN/SPC/144, 18/10/61 and *European Accelerator Study Group, second meeting,* Harwell, 30–31/10/61 (CERN archives, DG 20780).
17. Documents CERN/AR/int. SR/61-29, 27/11/61 and CERN/428 (CERN archives).
18. UK *DSIR and NIRNS, op. cit.* [Note 15]; CERN/442, 4/6/62.
19. As example, see what Bannier, ex-chairman of the Finance Committee of the Council said in CERN/FC/492, 21/7/61, p. 6.
20. For a critical analysis, Sfez (1981).
21. Schilling (1961).
22. Rudwick (1985), p. 451.
23. Veyne (1979); Furet (1985).
24. On those aspects, Pestre and Krige (1986). A longer bibliography is there.

Collected Bibliography

Bibliographical Note

All the notes and references following each chapter normally refer to this bibliography by author and year of publication. Entries from the *Dictionary of Scientific Biography* (*DSB*), anonymous articles and published material cited by institutions have been retained within the notes and references.

Manuscripts are cited only within individual chapters (where the necessary abbreviations will be found) and, for this purpose, internal laboratory reports are treated as manuscripts.

Accum, F. (1817). *Descriptive Catalogue of Apparatus and Instruments Employed in Operative Chemistry...* (1819), London: *Chemical Amusements*, 4th edn, London

Aikin, A. and Aikin, C.R. (1807). *Dictionary of Chemistry and Mineralogy*, 2 volumes, London

Allam, E.P. (1952). 'Reminiscences of Finsbury Technical College', *Central*, 17–18

Amato, G. *et al.* (1963). 'One-stage separated K-meson beam of 1.5 GeV/c momentum at the CPS', *Nuclear Instruments Meth.*, **20**, 47–50

Ampère, A.M. (1822). 'Expériences relatives à de nouveaux phénomènes électrodynamiques (1)', *Ann. Chim.*, **20**, 60–74

Anderson, R.G.W. (1978). *The Playfair Collection and the Teaching of Chemistry at the University of Edinburgh 1713–1858*, Edinburgh

Aris, R. *et al.* (1983). *Springs of Scientific Creativity*, Minneapolis

Armstrong, E.F. (1938). 'The City & Guilds of London Institute. Its origins and development', *Central*, **35**, 14–46

Armstrong, H.E. (1916). 'Personal notes on the origin and development of the chemical school at The Central', *Central*, **13**, 84–95

——(1934). 'The beginnings of Finsbury and the Central', *Central*, **31**, 1–15

Aronovitch, L. (1983). *Towards a New Knowledge of Nature: Physics at Harvard University, 1870–1910*, Harvard College AB honours thesis

Aubert, B. *et al.* (1963). 'Low energy separated beam at the CERN PS', *Nuclear Instruments Meth.*, **20**, 51–4

Auerbach, F. (1922). *Ernst Abbe: Sein Leben, sein Wiken, Seine Persönlichkeit*, 2nd edn, Leipzig

Auerbach, L. (1965). 'Scientists in the New Deal: A pre-war episode in the relations between science and government in the United States', *Minerva*, **3**, 457–82

Ayrton, W.E. (1892). 'Electrotechnics', *J. Inst. Elec. Eng.*, **21**, 5–36

Baddeley, W. (1835). 'Ede's portable chemical laboratory', *Mech. Mag.*, **23**, 163

Baggett, N.V. (Ed.) (1980). *AGS 20th Anniversary Celebration, May 22, 1980*, Brookhaven

Baltay, C. *et al.* (1961). 'Design and performance of a 3.3 BeV/c separated antiproton beam', in Blewett (1961), pp. 452–8

——(1963): 'The separated beam at the AGS — performance with antiprotons and π^+ mesons', *Nuclear Instruments Meth.*, **20**, 37–41

Barbier, M. *et al.* (1959). 'Studies of an experimental beam-stacking electron accelerator', *Int. Conf. High Energy Accel. Instrumentation*, Geneva, pp. 100–14

Batens, D. and van Bendegem, J.P. (Eds.) (1988). *Theory and Experiment*, Dordrecht and Boston

Batten, A.H. (1977). 'The Struves of Pulkowa — a family of astronomers', *J. Roy. Ast. Soc. Can.*, **71**, 345–72

Beddoes, T. (Ed.) (1799). *Contributions to Physical and Medical Knowledge, Principally from the West of England, Collected by Thomas Beddoes, M.D.*, Bristol

Belloni, L. (1970). 'The repetition of experiments and observations: its value in studying the history of medicine (and science)', *J. Hist. Med.*, **25**, 158–67

Bence Jones, H. (1862). *Report on the Past, Present and Future of the Royal Institution, Chiefly in Regard to Its Encouragement of Scientific Research*, London

——(1871). *The Royal Institution: Its Founder and Its First Professors*, London

Berman, M. (1978). *Social Change and Scientific Organization*, London and Ithaca

Beranek, L.L. (1977). 'The notebooks of Wallace C. Sabine', *J. Acoust. Soc. Am.*, **61**, 629–39

Bernstein, J. (1987). *The Life It Brings: One Physicist's Beginnings*, New York

Bessel, F.W. (1818). *Fundamenta Astronomiae*, Königsberg

——(1830). *Tabulae Regiomontanae*, Königsberg

——(1848). *Populäre Vorlesungen*, Hamburg

Black, J. (1803). *Lectures on the Elements of Chemistry* (edited by J. Robison), London and Edinburgh

Blewett, L. (forthcoming). 'Accelerator design and construction in the 1950s', in Brown *et al.* (forthcoming)

Blewett, M.H. (Ed.) (1961). *Int. Conf. High Energy Accel.*, 6–12 September, Brookhaven

Brachner, A. (1985). 'German nineteenth century instrument makers', in Clercq (1985), pp. 117–58

Brande, W.T. (1848). *A Manual of Chemistry*, London

Bridgman, P.W. (1941). 'Edwin Herbert Hall', *Biogr. Mem. Natl Acad. Sci.*, **21**

Brock, W.H. (1973). *H.E. Armstrong and the Teaching of Science 1889–1930*, Cambridge

——(1976). 'The spectrum of science patronage', in Turner (1976), pp. 173–206

——(1980). 'Observe, experiment and conclude. Finsbury College's new course of experimental philosophy in 1879–1880', *Hist. Ed. Soc. Bull.*, **26**, 46–50

——(1981). 'The Japanese connexion. Engineering in Tokyo, London and Glasgow at the end of the 19th century', *Br. J. Hist. Sci.*, **14**, 227–43

——and Price, M.H. (1980). 'Squared paper in the nineteenth century: Instrument of science and engineering, and symbol of reform in mathematical education', *Ed. Stud. Math.*, **11**, 365–81

Brougham, H. (1825). *Practical Observations on the Education of the People*, London

——(1826). 'Objects, advantages, and pleasures of science', *Pamphleteer*, **27**; reprinted in Society for the Diffusion of Useful Knowledge, *Library of Useful Knowledge*, London, pp. 1–48

——(1855). *Lives of Philosophers of the Time of George III*, London and Glasgow

Brown, L., Dresden, M. and Hoddeson, L. (Eds.) (forthcoming). *Pions to Quarks: History of Particle Physics in the 1950s*, Cambridge

Browne, C.A. (1925). 'The life and chemical services of Frederick Accum', *J. Chem. Ed.*, **2**, 1–27

Bryant, M. (1986). *The London experience of secondary education*, London and New Jersey

Buchanan, P. and Hunt, L.B. (1984). 'Richard Knight (1768–1844): A forgotten chemist and apparatus dealer', *Ambix*, **31**, 57–67

Buchwald, J.Z. (1985). *From Maxwell to Microphysics*, Chicago

Bud, R. and Roberts, G.K. (1985). *Science versus Practice*, Manchester

Cahan, D. (1982). 'Werner Siemens and the origin of the Physikalisch-Technische Reichsanstalt, 1872–1887', *Hist. Stud. Phys. Sci.*, **12**, 253–83

——(1985). 'The institutional revolution in German physics, 1865–1914', *Hist. Stud. Phys. Sci.*, **15**, 1–65

——(forthcoming). *An Institute for an Empire: The Physikalisch-Technische Reichsanstalt, 1871–1918*, Cambridge and New York

Cantor, G.N. (1989). 'The rhetoric of experiment', in Gooding *et al.* (1989), pp. 159–80

Carhart, H. (1900). 'The Imperial Physico-Technical Institution in Charlottenburg', *Trans. Am. Inst. Elec. Eng.*, **17**, 555–83

Cary, W. (1827). *Catalogue of Optical, Mathematical, and Philosophical Instruments Made and Sold by W. Cary, no. 182 Strand, Near Norfolk Street*, London

Cattell, J.M. (1910). 'A further statistical study of American men of science', *Science, N.Y.*, new series, **32**, 633–48

Cattermole, M.J.E. and Wolfe, A.F. (1987). *Horace Darwin's Shop. A History of the Cambridge Scientific Instrument Company 1878 to 1968*, Bristol

Children, J.G. (1809–15). 'An account of some experiments with a large voltaic battery', *Phil. Trans*, **99**, 32–8; **105**, 363–74

Christison, R. (1885–6). *The Life of Sir Robert Christison, Bart.*, 2 volumes, Edinburgh and London

Cini, M. (1980). 'The history and ideology of dispersion relations: The pattern of internal and external factors in a paradigm shift', *Fund. Sci.*, **1**, 157–72

Clay, F. (1902). *Modern School Buildings: Elementary and Secondary*, 1st edn, London (2nd edn, London, 1904; 3rd edn, London, 1921)

Clercq, P.R. de (1985). *Nineteenth-century Scientific Instruments and Their Makers*, Leiden and Amsterdam

Cocconi, G. (1960). 'Progress report on work with the 25 GeV proton synchrotron', *Proc. 1960 Ann. Int. Conf. High Energy Phys.*, 25 August–1 September, Rochester

Cochrane, T. (1966). *Notes from Doctor Black's Lectures on Chemistry 1767/8* (ed. D. McKie), Wilmslow, Cheshire

Coker, E.G. (1907). 'The new engineering laboratory . . .', *Engineering*, 16 August, 232–3

Cole, F. (1971). 'Progress report on the NAL accelerator', *Part. Acc.*, **2**, 1–11

Coleby, L. J. M. (1952). 'John Hadley, Fourth Professor of Chemistry in the University of Cambridge', *Ann. Sci.*, **8**, 293–301

——(1953). 'Richard Watson, Professor of Chemistry in the University of Cambridge', *Ann. Sci.*, **9**, 101–23

Collins, H.M. (1974). 'The TEA set: Tacit knowledge and scientific net works', *Sci. Stud.*, **4**, 165–85

——(Ed.) (1981). 'Knowledge and controversy: Studies of modern natural sciences', *Soc. Stud. Sci.*, **11**, 1–158

Comrie, J.D. (1939). 'Boerhaave and the early Medical School at Edinburgh', in *Memorialia Herman Boerhaave Optimi Medici*, Haarlem

Courant, E.D. (forthcoming). 'Early history of the cosmotron and AGS at Brookhaven', in Brown *et al.* (forthcoming)

—— and Cool, R. (1959). 'Transport and separation of beams from AG synchrotron', in Kowarski (1959), pp. 403–12

Coutts, J. (1909). *A History of the University of Glasgow from Its Foundation in 1451 to 1909*, Glasgow

Cresswell, H.B. (1975). 'Sir Aston Webb and his office', in *Service* (1975), 328–37

Crönstedt, A.F. (1770). *An Essay towards a System of Mineralogy*, London

Crowther, J.G. (1974). *The Cavendish Laboratory 1874–1974*, Cambridge

Crozon, M. (1987). *La matière première, Le Recherche des particules fondamentals et de leurs interactions*, Paris

Cullen, W. (1756). 'Of the cold produced by evaporating fluids, and some other means of producing cold', *Essays and Observations, Physical and Literary*, **2**, 145–56

Daniels, G.H. (1967). 'The process of professionalization in American science: The emergent period, 1820–1860', *ISIS*, **58**, 151–66

——(1968). *American Science in the Age of Jackson*, New York

——(Ed.) (1972). *Nineteenth-century American Science: A Reappraisal*, Evanston

Davy, H. (1799a). 'An essay on heat, light and the combinations of light', in Beddoes (1799), pp. 4–147. Addenda on pp. 199–205

——(1799b). 'An essay on the generation of phosoxygen, or oxygen gas, and on the causes of the colours of organic beings', in Beddoes (1799), pp. 151–98. Addenda on pp. 199–205

——(1810). *A Lecture on the Plan which it is Proposed to Adopt for Improving the Royal Institution, and Rendering it Permanent*, London. Reprinted in *Proc. Roy. Inst.* (1914–16), **21**, pp. 1–16

——(1821a). 'On the magnetic phenomena produced by electricity', *Phil. Trans.*, **111**, 7–19

——(1821b). 'Farther researches on the magnetic phenomena produced by electricity with some new experiments on the properties of electrified bodies in their relations to conducting powers and temperature', *Phil. Trans.*, **111**, 425–39

Davy, J. (1836). *Memoirs of the Life of Sir Humphry Davy*, 2 volumes, London

——(Ed.) (1839–40). *The Collected Works of Sir Humphry Davy*, 9 volumes, London

Denke, R. (1956). 'Das Siemens-Grundstück: Ein Beitrag zu seiner Geschichte', *Der Bär von Berlin: Jahrbuch des Vereines für die Geschichte Berlins*, **6**, 108–34

Donovan, A. (1979). 'Scottish responses to the new chemistry of Lavoisier', *Stud. 18th Cent. Cult.*, **9**, 237–49

Durm, J., Ende, H., Schmitt, E. and Wagner, H. (Eds.) (1888). *Handbuch der Architektur*, Teil 4: *Entwerfen, Anlagen und Einrichtung der Gebäude*, Halbband 6: *Gebäude für Erziehung, Wissenschaft und Kunst*, Heft 2: *Hochschulen, zugehörige und verwandte wissenschaftliche Institut*, Darmstadt

Ede, R.B. (1837). *Practical Facts of Chemistry*, London

——(1843). *Practical Facts of Chemistry*, 2nd edn, London

Edgeworth, M. and Edgeworth, R.L. (1798). *Practical Education*, 2 volumes, London

Eliot, C.W. (1869). 'The new education', *The Atlantic Monthly*, **23**, 203–22, 358–67

Ellul, J. (1964). *The Technological Society*, New York

England, J.M. (1982). *A Patron for Pure Science: The National Science Foundation's Formative Years 1945–57*, Washington

Eve, A.S. (1939). *Rutherford*, Cambridge

Eyre, J.V. (1958). *Henry Edward Armstrong 1848–1937*, London

Falconer, I.J. (1988). 'Thomson's work on positive rays 1906–1914', *Hist. Stud. Phys. Sci.*, **18**, 265–310

Faraday, M. (1821). 'On some new electromagnetical motions, and on the theory of magnetism', *Quart. J. Sci.*, **12**, 74–96

——(1821–2). 'Historical sketch of electro-magnetism', *Ann. Phil.*, **18**, 195–200, 274–90; **19**, 107–21

——(1827). *Chemical Manipulation*, London

Farrar, W.V., Farrar, K.R. and Scott, E.L. (1976). 'The Henrys of Manchester. Part 5', *Ambix*, **23**, 27–52

Fawcett, J. (Ed.) (1976). *Seven Victorian Architects*, London

Fenby, D.V. (1987). 'Heat. Its measurement from Galileo to Lavoisier', *Pure App. Chem.*, **59**, 91–100

Ferguson, E.S. (1977). 'The mind's eye: Nonverbal thought in technology', *Science, N.Y.*, **197**, 827–36

Finn, B.S. (1971). 'Output of 18th-century electrostatic machines', *Br. J. Hist. Sci.*, **5**, 289–91

Fleck, L. (1979). *Genesis and Development of a Scientific Fact*, Chicago and London

Fleming, D.R. (1965). 'The big money and high politics of science', *Atlantic Monthly*, **216**, 41–5

Foden, F. (1970). *Philip Magnus, Victorian Educational Pioneer*, London

Forbes, E.G. (1975). *Greenwich Observatory: Its Origins and Early History*, London

Forgan, S. (1986). 'Context, image and function: a preliminary enquiry into the architecture of scientific societies', *Br. J. Hist. Sci.*, **19**, 89–113

Forman, P. (1971). 'Weimar culture, causality, and quantum theory, 1918–1927: Adaption by German physicists to a hostile intellectual milieu', *Hist. Stud. Phys. Sci.*, **3**, 1–115

——(1987). 'Behind quantum electronics: National security as basis for physics research in the United States, 1940–1960', *Hist. Stud. Phys. Biol. Sci.*, **18**, 149–229

——, Heilbron, J. and Weart, S. (1975). 'Physics *circa* 1900: Personnel, funding, and productivity of the academic establishments', *Hist. Stud. Phys. Sci.*, **5**, 1–185

Franklin, A. (1979). 'The discovery and nondiscovery of parity nonconservation', *Stud. Hist. Phil. Sci.*, **10**, 201–57

French, J.C. (1946). *A History of the University Founded by Johns Hopkins*, Baltimore

Fullmer, J.Z. (1967). 'Davy's sketches of his contemporaries', *Chymia*, **12**, 127–50

Furet, F. (1985). *Penser la Révolution Française*, Paris

Galison, P. (1983). 'The discovery of the muon and the failed revolution against quantum electrodynamics', *Centaurus*, **26**, 262–316

——(1985). 'Bubble chambers and the experimental workplace', in Hannaway and Achinstein (1985), pp. 309–73

——(1987). *How Experiments End*, Chicago and London

——(forthcoming). 'Physics between war and peace', in Mendelsohn and Smith (forthcoming)

——and Assmus, A. (1989). 'Artificial clouds, real particles', in Gooding *et al.* (1989), pp. 225–74

Galton, D. (1895): 'On the Reichsanstalt, Charlottenburg, Berlin', *Rep. Br. Ass.*, 606–8

Galton, F. (1874). *English Men of Science: Their Nature and Nurture*, London

Garren, A., Lambertson, G., Lofgren, E. and Smith, L. (1967). 'Extendible-energy synchrotron', *Nuclear Instruments Meth.*, **54**, 223

Geison, G.L. and Secord, J.A. (1988). 'Pasteur and the process of discovery: The case of optical isomerism', *ISIS*, **79**, 7–36

Gell-Mann, M., Goldberger, M.L. and Thirring, W.E. (1954). 'Use of causality conditions in quantum theory', *Phys. Rev.*, **95**, 1612–27

Germain, C. and Tinguely, R. (1963). 'Electrostatic separator technique at CERN', *Nuclear Instruments Meth.*, **20**, 21–5

Gibbs, F.W. (1951a). 'Robert Dossie (1717–1777) and the Society of Arts', *Ann. Sci.*, **7**, 149–72

——(1951b). 'Peter Shaw and the revival of chemistry', *Ann. Sci.*, **7**, 221–37

——(1952a). 'William Lewis, MB, FRS', *Ann. Sci.*, **8**, 122–51

——(1952b). 'Essay review: Prelude to chemistry in industry', *Ann. Sci.*, **8**, 271–81

Gibbs, W., Pickering, E.C. and Trowbridge, J. (1879). 'List of apparatus', *Harvard Coll. Lib. Bull.*, 302–4, 350–4

Gilman, D.C. (1906). *The Launching of a University*, New York

Godwin, G. (1854). *London Shadows: A Glance at the Homes of the Thousands*, London

——(1859). *Town Swamps and Social Bridges*, London, reprinted Leicester, 1972

Goldberger, M.L. (1955a). 'The use of causality conditions in quantum theory', *Phys. Rev.*, **97**, 508–10

——(1955b). 'Causality conditions and dispersion relations. I. Boson fields', *Phys. Rev.*, **99**, 979–85

——(1961). 'Theory and applications of single variable dispersion relations', in Stoops (1961), pp. 179–95

——(1970). 'Fifteen years in the life of dispersion relations', in Zichichi (1970), Vol. 2, pp. 684–93

Goodfield, J. (1982). *An Imagined World: A Story of Scientific Discovery*, Harmondsworth

Gooding, D. (1982). 'Empiricism in practice: Teleology, economy and observation in Faraday's physics', *ISIS*, **73**, 46–67

——(1985a). '"He who proves, discovers": John Herschel, William Pepys and the Faraday Effect', *Notes Rec. Roy. Soc. Lond.*, **39**, 229–44

——(1985b). '"In Nature's School": Faraday as an experimentalist', in Gooding and James (1985), pp. 105–35

——(1989a). '"Magnetic curves" and the magnetic field: Experimentation and representation in the history of a theory', in Gooding *et al.* (1989), pp. 182–223

——(1989b): *The Making of Meaning*, Dordrecht and Boston

——and James, F.A.J.L. (Eds.) (1985). *Faraday Rediscovered: Essays on the Life and Work of Michael Faraday, 1791–1867*, London and New York

——, Pinch, T. and Schaffer, S. (Eds.) (1989). *The Uses of Experiment: Studies in the Physical Sciences*, Cambridge

Görges, H. (Ed.) (1929). *Elektrotechnischer Verein: Festschrift...*, Berlin

Göttling, J.F.A. (1791). *Description of a Portable Chest of Chemistry*, London

Greenaway, F., Berman, M., Forgan, S. and Chilton, D. (Eds.) (1971–6). *Archives of the Royal Institution, Minutes of Managers' Meetings, 1799–1903*, 15 volumes, bound in 7, London

Greenbaum, L. (1971). *A Special Interest*, Ann Arbor

Greenberg, D.S. (1967). *The Politics of Pure Science*, New York

Griffin, J.J. (1825). *Chemical Recreations*, 5th edn, London

——(1838). *Chemical Recreations*, Glasgow

——(1849). *Chemical Recreations*, Glasgow

——(1858). *The Radical Theory of Chemistry*, London

Hacking, I. (1983). *Representing and Intervening: Introductory Topics in the Philosophy of Natural Science*, Cambridge

Hackmann, W.D. (1978). 'Eighteenth century electrostatic measuring devices', *Ann. Inst. Mus. Stor. Sci. Firenze*, **3**, 3–58

——(1979). 'The relationship between concept and instrument design in eighteenth-century experimental science', *Ann. Sci.*, **36**, 205–24

——(1985). 'Instrumentation in the theory and practice of science: Scientific instruments as evidence and as an aid to discovery', *Ann. Inst. Mus. Stor. Sci. Firenze*, **20**, 87–115

Haemmerling, K. (1955). *Charlottenburg: Das Lebensbild einer Stadt, 1905–1955*, Berlin

Haesslin, J.J. (Ed.) (1971). *Berlin*, Munich

Hagen, E. and Scheel, K. (1906). 'Die Physikalisch-Technische Reichsanstalt', *Ingenieurwerke in und bei Berlin: Festschrift zum 50 Jahrigen Bestehen des VDI*, Berlin, pp. 60–7

Hale, A.J. (1921–5). *Modern Chemistry, Pure and Applied*, 5 volumes, London

Hall, E.H. (1887). *Harvard Descriptive List of Elementary Physical Experiments*, Cambridge, Mass.

——(1903). 'Do falling bodies move south? I. Historical, II. Method and results of the author's work', *Phys. Rev.*, **17**, 179–90, 245–54

——(1904). 'Experiments of the deviations of falling bodies', *Proc. Am. Acad. Arts Sci.*, **39**, 339–49

——(1913). *College Laboratory Manual of Physics*, Cambridge, Mass.

——(1919). 'Benjamin Osgood Peirce', *Biogr. Mem. Natl Acad. Sci.*, **8**

——(1930). 'Physics', in Morison (1930)

——(1932). 'John Trowbridge', *Biogr. Mem. Natl Acad. Sci.*, **14**

——and Bergen, J.Y. (1895). *A Textbook of Physics*, New York

Hamilton, W. (Ed.) (1863). *The Works of Thomas Reid, D.D.*, 2 volumes, Edinburgh

Hannaway, O. and Achinstein, P. (Eds.) (1985). *Experiment and Observation in Modern Science*, Cambridge, Mass.

Harman, P.M. (Ed.) (1985). *Wranglers and Physicists: Studies on Cambridge [Mathematical] Physics in the Nineteenth Century*, Manchester

Harrison, A. (1978). *Making and Thinking: A Study of Intelligent Activities*, Brighton

Harte, N. (1986). *The University of London 1836–1986*, London, 1986

Hartley, H. (1960). 'Sir Humphrey Davy, Bt., P.R.S.', *Proc. Roy. Soc.*, **255A**, 161–2

——(1966). *Humphry Davy*, London

Hawkins, H. (1960). *Pioneer: A History of the Johns Hopkins University*, Ithaca

Hays, J.N. (1983). 'The London lecturing empire', in Inkster and Morrell (1983), pp. 91–119

Heilbron, J.L., Seidel, R.W. and Wheaton, B.R. (1981). *Lawrence and His Laboratory: Nuclear Science at Berkeley, 1931–1961*, Berkeley

Hendry, J. (Ed.) (1984). *Cambridge Physics in the Thirties*, Bristol

Henry, W. (1806). *Epitome of Experimental Chemistry*, 4th edn, Edinburgh

——(1810). *Elements of Chemistry*, 6th edn, London

Hermann, A., Krige, J., Mersits, U. and Pestre, D. (1987 and forthcoming). *History of CERN*, 2 volumes, Amsterdam

Herschel, J.F.W. (1830). *A Preliminary Discourse on the Study of Natural Philosophy*, London

Hertz, H. (1927). *Erinnerungen, Briefe, Tagebücher* (ed. Hertz, J.), Leipzig

Hetherington, N.S. (1976). 'Cleveland Abbe and a view of science in a mid-nineteenth century America', *Ann. Sci.*, **33**, 31–49

Hewlett, R. and Anderson, O. (1962). *The New World, 1939/1946: A History of the United States Atomic Energy Commission, Vol. I*, University Park, Pennsylvania

Heydweiller, A. (1910–11). 'Friedrich Kohlrausch', in Kohlrausch (1910–11), Vol. 2, pp. xxxv–lxxii

Hill, K. (Ed.) (1964). *The Management of Science*, Boston

Hoddeson, L. (1983). 'Establishing KEK in Japan and Fermilab in the US: Internationalism, nationalism and high energy accelerator physics during the 1960s', *Soc. Stud. Sci.*, **13**, 1–48

——(1987). 'The first large-scale application of superconductivity: The Fermilab energy doubler, 1972–1983', *Hist. Stud. Phys. Biol. Sci.*, **18**, 25–54

Holborn, L. (1895). 'Elektrische Strassenbahnen und physikalische Institut', *Preuss, Jahr.*, **81**, 177–84

Holmes, F.L. (1985). *Lavoisier and the Chemistry of Life: An Exploration of Scientific Creativity*, Madison

——(1987). 'Scientific writing and scientific discovery', *ISIS*, **78**, 220–35

Holton, G. (1984). 'How the Jefferson Physical Laboratory came to be', *Phys. Today*, **37**, 32–7

Inkster, I. and Morrell, J.B. (Eds.) (1983). *Science in the Metropolis: Science in British Culture, 1780–1850*, London

Ironmonger, E. (1958). 'The Royal Institution and the teaching of science in the nineteenth century', *Proc. Roy. Inst.*, **37**, 139–58

Irvine, J.H. and Martin, B.R. (1984). 'CERN: Past performance and future prospects — II. The scientific performance of the CERN accelerators', *Res. Pol.*, **13**, 247–84

Jachim, A. (1975). *Science Policymaking in the United States and the Batavia Accelerator*, Carbondale

Jaffe, G. (1952). 'Recollections of three great laboratories', *J. Chem. Ed.*, **29**, 230–8

Jenkins, E.W. (1979). *From Armstrong to Nuffield*, London

Jenkins, R.V. (1987). 'Words, images, artifacts and sound: Documents for the history of technology', *Br. J. Hist. Sci.*, **20**, 39–56

Jordan, D.W. (1985). 'The cry for useless knowledge: Education for a new Victorian technology', *Inst. Elec. Eng. Proc.*, **A132**, 587–601

Jorpes, E.J. (1966). *Jac. Berzelius. His Life and Work*, Stockholm

Junk, C. (1905). 'Physikalische Institut', in Schmitt *et al.* (1905), pp. 164–236

Kearns, D. (1976). *Lyndon Johnson and the American Dream*, New York

Kemble, E.C. and Birch, F. (1970). 'Percy Williams Bridgman', *Biogr. Mem. Natl Acad. Sci.*, **41**

Kent, A. (Ed.) (1950). *An Eighteenth Century Lectureship in Chemistry*, Glasgow

Kerst, W. (forthcoming). 'Accelerator developments of the Midwestern Universities Research Association (MURA) in the 1950s', in Brown *et al.* (forthcoming)

Kevles, D.J. (1978). *The Physicists: The History of a Scientific Community in Modern America*, New York

King, H.C. (1955). *The History of the Telescope*, London

Knorr-Cetina, K.D. (1981). *The Manufacture of Knowledge: An Essay on the Constructivist and Contextual Nature of Science*, Oxford

——and Mulkay, M. (Eds.) (1983). *Science Observed*, Beverly Hills and London

Kohlrausch, F. (1895). 'Diskussion über die Frage der Störungen wissenschaftlicher Institut durch electrische Bahnen', *Elektrotech. Zeit.*, **16**, 427–9, 444–5

——(1910–11). *Gesammelte Abhandlungen*, 2 volumes (ed. Hallswachs, W., Heydweiller, A., Strecker, K. and Wiener, O.), Leipzig

Kohlstedt, S.G. (1976). *The Formation of the American Scientific Community*, Urbana

Kowarski, L. (Ed.) (1959). *Proc. Int. Conf. High-Energy Accel. Instrumentation — CERN 1959*, 14–19 September, Geneva

Krebs, H. (with Schmid, R.) (1979). *Otto Warburg: Zellphysiologe, Biochemiker, Mediziner, 1883–1970*, Stuttgart

Krige, J. and Pestre, D. (1986). 'The choice of CERN's first large bubble chambers for the proton synchrotron (1957–1958)', *Hist. Stud. Phys. Biol. Sci.*, **16**, 255–79

Krisciunas, K. (1988). *Astronomical Centres of the World*, Cambridge

Lang, J. (1978). *City and Guilds of London Institute. An Historical Commentary*, London

Lang, M. (1978). 'Maria Edgeworth's *The Parent's Assistant*', *Hist. Ed.*, **7**, 21–33

Larmor, J. (Ed.) (1907). *Memoir and Scientific Correspondence of the Late Sir George Gabriel Stokes*, 2 volumes, Cambridge

Latour, B. (1983). 'Give me a laboratory and I will raise the world', in Knorr-Cetina and Mulkay (1983), pp. 141–70

——and Woolgar, S. (1979). *Laboratory Life: The Construction of Scientific Facts*, Beverly Hills and London; 2nd edn, Princeton, 1986

Layton, D. (1975). 'Science or education', *Univ. Leeds Rev.*, No. 18

Lehmann, O. (1911). *Geschichte des Physikalischen Instituts der Technischen Hochschule Karlsruhe: Festgabe der Fridericiana zur 83. Versammlung Deutscher Naturforscher und Ärzte*, Karlsruhe

Leitner, J. *et al.* (1963). 'Performance of the AGS separated beam with high energy kaons', *Nuclear Instruments Meth.*, **20**, 42–6

Leslie, S.W. (1987). 'Playing the education game to win: The military and interdisciplinary research at Stanford', *Hist. Stud. Phys. Biol. Sci.*, **18**, 55–88

Lewis, W. (1763). *Commercium Philosophico-technicum; or, The Philosophical Commerce of Arts*, London

Leyden, F. (1933). *Gross-Berlin: Geographie der Weltstadt*, Breslau

——(1971). 'Charlottenburg um 1880', in Haesslin (1971), pp. 322–4

Lindeboom, G.A. (1968). *Hermann Boerhaave. The Man and his Work*, London

——(1974). *Boerhaave and Great Britain*, Leiden

Livingston, M.S. (1968a). *Early History of the 200-GeV Accelerator*, Batavia

——(1968b). *Design Progress at the National Accelerator Laboratory: 1968–1969*, Batavia

——(1969). *Particle Accelerators: A Brief History*, Cambridge

——and Blewett, J.P. (1962). *Particle Accelerators*, New York

Lofgren, E.J. (1956). 'Bevatron operational experiences', *Proc. CERN Symp. High Energy Accel. Pion Phys.*, 11–23 June, Vol. 1, p. 500

Lowe, R.A. and Knight, R. (1982). 'Building the ivory tower: the social functions of late 19th-century collegiate architecture', *Stud. Higher Ed.*, **7**, 81–91

Lowi, T.J. and Ginsberg, B. (1976). *Poliscide*, New York

Lyman, T. (1914): 'An extension of the spectrum in the extreme ultra-violet', *Nature, Lond.*, **93**, 241

[——]* (1932). *The Physical Laboratories of Harvard University*, Cambridge, Mass.

McMillan, E. (1984). 'A history of the synchrotron', *Phys. Today*, **37**, 31–7

Magnus, P. (1883). *Technical Instruction. The introductory address . . . at the opening of the Finsbury Technical College*, London

——(1910). *Educational Aims and Efforts 1880–1910*, London

Marchant, J. (Ed.) (1916). *Raphael Meldola*, London

Martin, T. (Ed.) (1932–6). *Faraday's Diary. Being the Various Philosophical Notes of Experimental Investigation Made by Michael Faraday, DCL, FRS, during the Years 1820–1862 . . .* , 7 volumes and index, London

Medawar, P. (1964). 'Is the scientific paper a fraud?', *Saturday Rev.*, 1 August, 43–4

Mendelsohn, E. (1964). 'The emergence of science as a profession in nineteenth-century Europe', in Hill (1964), pp. 3–48

*Written anonymously at the time.

—— and Smith, M.R. (forthcoming). *Science, Technology and the Military. Sociology of the Sciences Yearbook, 1987*, Boston

Miller, J.D. (1970). *Henry Augustus Rowland and His Electromagnetic Researches*, Oregon State University PhD thesis

Millis, C.T. (1925). *Technical Education. Its Development and Aims*, London

Morison, S.E. (Ed.) (1930). *The Development of Harvard University*, Cambridge, Mass.

——(1936). *Three Centuries of Harvard*, Cambridge, Mass.

Morley, J. (1971). *Death, Heaven and the Victorians*, London

Morrell, J.B. (1969a). 'Thomas Thomson: Professor of chemistry and university reformer', *Brit. J. Hist. Sci.*, **4**, 245–65

——(1969b). 'Practical chemistry in the University of Edinburgh', *Ambix*, **16**, 66–80

——(1972). 'The chemist breeders: The research schools of Liebig and Thomas Thomson', *Ambix*, **19**, 1–46

Müller, K.A.v. (Ed.) (1926). *Die wissenschaften Anstalten der Ludwig-Maximillians-Universität zu München: Chronik zur Jahrhundertfeier*, Munich

Munby, A.E. (1929). *School Laboratory Fittings*, London

Naglo, E. (Ed.) (1904). *Die ersten 25 Jahre des Elektrotechnischen Vereins*, Berlin

Naylor, R. (1989). 'Galileo's experimental discourse', in Gooding *et al.* (1988), pp. 116–34

Needell, A.S. (1983): 'Nuclear reactors and the founding of Brookhaven National Laboratory', *Hist. Stud. Phys. Sci.*, **14**, 93–122

Newcomb, S. (1903). *Reminiscences of an Astronomer*, Boston and New York

Newman, J. (1827). *A Catalogue of Optical, Mathematical, and Philosophical Instruments Manufactured and Sold by John Newman, 122 Regent Street*, London

Nicholson, W. (1795). *Dictionary of Chemistry*, 2 volumes, London

Nickles, T. (1984). 'Justification as discoverability, II', *Philosophia Naturalis*, **21**, 563–76

——(1985). 'Beyond divorce: Current status of the discovery debate', *Phil. Sci.*, **52**, 177–207

——(1988). 'Reconstructing science: Discovery and experiment', in Batens and van Bendegem (1988), pp. 33–53

——(1989). 'Justification and experiment', in Gooding *et al.* (1989), pp. 299–333

Norton, M., Katzman, D., Escot, P., Chudacoff, H., Paterson, T. and Tuttle, W. (Eds.) (1982). *A People and a Nation: A History of the United States, Volume II*, Dallas

Owen, D. (1964). *English Philanthropy 1660–1960*, Cambridge

Pais, A. (1986). *Inward Bound: Of Matter and Forces in the Physical World*, Oxford

Paris, J.A. (1831). *The Life of Sir Humphry Davy*, 2 volumes, London

Parkes, S. (1822). *Chemical Catechism*, London

Peirce, B.O. (1909). 'Joseph Lovering', *Biogr. Mem. Natl Acad. Sci.*, **6**, 335–7

Pepys, W.H. (1823). 'An account of an apparatus on a peculiar construction for performing electro-magnetic experiments, *Phil. Trans.*, **113**, 187–8

Pernet, J. (1891). 'Über die physikalisch-technische Reichsanstalt zu Charlottenburg und die daselbst ausgeführten elektrischen Arbeiten', *Schweizer. Bauzeit.*, **18**, 1–6

Pestre, D. (1988). 'Comment se prennent les décisions de très gros équipements dans les laboratoires de 'science lourde' contemporains, un récit suivi de commentaires', *Rev. Syn.*, **4**, 97–130

—— and Krige, J. (1986). 'Le comment et le pourquoi de la naissance du CERN', *Relat. Int.*, **46**, 209–26

Phillips, D. (1979). 'William Lawrence Bragg', *Biogr. Mem. Fell. Roy. Soc.*, **25**, 75–143

Phillips, M. (1983). 'Laboratories and the rise of physics in the nineteenth century', *Am. J. Phys.*, **51**, 497–503

Pickering, A. (1984). *Constructing Quarks: A Sociological History of Particle Physics*, Chicago and Edinburgh

——(1989). 'Living in the material world', in Gooding *et al.* (1989), pp. 275–97

——(forthcoming). 'From field theory to phenomenology: The history of dispersion relations', in Brown *et al.* (forthcoming)

—— and Trower, W.P. (1985). 'Sociological problems of high-energy physics', *Nature, Lond.*, **318**, 243–5

Pingree, J. (n.d.). *List of the Papers and Correspondence of Henry Edward Armstrong*, London
—— (1974). *Armstrong Papers. Second series*, London
Polanyi, M. (1964). *Personal Knowledge*, New York
Port, M.H. (Ed.) (1976). *The Houses of Parliament*, London
Poynter, J.R. (1969). *Society and Pauperism*, London
Price, D.J. de S. (1980). 'Philosophical mechanism and mechanical philosophy: Some notes towards a philosophy of scientific instruments', *Ann. Inst. Mus. Stor. Sci. Firenze*, **5**, 75–85
Price, M.P. (1983). 'Mathematics in English education 1860–1914: Some questions and explanations in curriculum history', *Hist. Ed.*, **12**, 271–84
Pusey, N. (Ed.) (1969). *A Turning Point in Higher Education*, Cambridge, Mass.
Ramm, C.A. (1960). 'Some features of beam-handling equipment for the CERN proton synchrotron', *Proc. Int. Conf. Instrumentation High-Energy Phys.*, 12–14 September, Berkeley, pp. 289–98
Ramsay, W. (1918). *The Life and Letters of Joseph Black, M.D.*, London
Ramsey, N. (1966). *Early History of Associated Universities and Brookhaven National Laboratory*, Brookhaven
Rayleigh, 3rd Lord (1899–1920). *Scientific Papers*, 6 volumes, Cambridge
Rayleigh, 4th Lord (1942). *The Life of Sir J.J. Thomson*, Cambridge
Reagan, M.D. (1969). *Science and the Federal Patron*, New York
Redlich, F. (1971). 'Science and charity: Count Rumford and his followers', *Int. Rev. Soc. Hist.*, **16**, 184–216
Reid, D.B. (1835a). *Textbook for Students of Chemistry*, Edinburgh
——(1835b). 'On the extension of the study of physics', *Rep. Brit. Ass.*, 126
——(1844). *Illustrations of the Theory and Practice of Ventilation*, London
Reingold, N. (1964). 'Cleveland Abbe at Pulkowa: Theory and practice in the nineteenth century physical sciences', *Arch. Int. Hist. Sci.*, **17**, 133–47
——(1972). 'American indifference to basic research: A reappraisal', in Daniels (1972)
Roberts, G.K. (1976). 'The establishment of the Royal College of Chemistry: An investigation of the social context of early Victorian chemistry', *Hist. Stud. Phys. Sci.*, **7**, 437–85
Robins, E.C. (1881). 'On the revelations of sanitary science', *Sanitary Record*, 15 October
——(1882–3). 'Building for applied science and art instruction', *Trans. Roy. Inst. Brit. Arch.*, 99–100
——(1885a). *Papers on Technical Education, Applied to Science Buildings, Fittings and Sanitation*, London
——(1885b). 'The ventilation and warming of chemical laboratories', *Trans. Sanitary Inst.*, **7**, 12
——(1887). *Technical School and College Buildings, Being a Treatise on the Design and Construction of Applied Science and Art Buildings, and Their Suitable Fittings and Sanitation, with a Chapter on Technical Education*, London
Robinson, H.R. (1954). 'Rutherford: Life and work to the year 1919, with personal reminiscences of the Cambridge period', in *Rutherford by Those Who Knew Him*, London, pp. 1–21
Rochester, C.D. (forthcoming). 'Cosmic ray cloud chamber contributions to the discovery of strange particles in the decade 1947–1957', in Brown *et al.* (forthcoming)
—— and Butler, C.C. (1947). 'Evidence for the existence of new unstable elementary particles', *Nature, Lond.*, **160**, 855
Rowland, H.A. (1883). 'A plea for pure science', in Rowland (1902), p. 597
——(1902). *Physical Papers*, Baltimore
Rudwick, M.J.S. (1985). *The Great Devonian Controversy*, Chicago
Russell, T.H. (1903). *Planning and Fitting of Chemical and Physical Laboratories*, London
Sands, M. (1960). *A Proton Synchrotron for 300 GeV*, Pasadena
Sanford, J. (1976). 'The Fermi National Accelerator', *Ann. Rev. Nuclear Sci.*, **26**, 165
Schaffer, S. (1986). 'Scientific discoveries and natural philosophy', *Soc. Stud. Sci.*, **16**, 387–420
——(1989). 'Glass works: Newton's prisms and the uses of experiment', in Gooding *et al.* (1989), pp. 67–104
Scherer, F.M. (1965). 'Invention and innovation in the Watt–Boulton steam-engine venture', *Tech. Cult.*, **6**, 165–87

Schiff, L.I. (1946). 'Production of particle energies beyond 200 MeV', *Rev. Sci. Instrum.*, **17**, 6–14

Schilling, W.R. (1961). 'The H-bomb decision: How to decide without actually choosing', *Pol. Sci. Quart.*, **76**, 24–46

Schmitt, E., Durm, J. and Ende, H. (Eds.) (1905). *Handbuch der Architektur, Teil 4: Entwerfen, Anlagen und Einrichtung der Gebäude*, Halbband 6: *Gebäude für Erziehung, Wissenschaft und Kunst*, Heft 2, a: *Hochschulen, zugehorige und verwandte wissenschaftliche Institut*, I: *Universitäten und Technische Hochschulen. Physikalische und chemische Institut. Mineralogische und geologische, botanische und zoologische Institut*, 2nd edn, Stuttgart

Schofield, R.E. (1963). *The Lunar Society of Birmingham*, Oxford

Schumpeter, J.A. (1934). *The Theory of Economic Development: An Inquiry into Profits, Capital, Credit, Interest, and the Business Cycle*, Cambridge

Schuster, A. (1884). 'Experiments on the discharge of electricity through gases: Sketch of a theory', *Proc. Roy. Soc.*, **37A**, 317–39

——(1911). *The Progress of Physics During 33 Years*, Cambridge

—— and Shipley, A. (1917). *Britain's Heritage of Science*, London

Schweber, S.S. (1986). 'The empiricist temper regnant: Theoretical physics in the United States 1920–1950', *Hist. Stud. Phys. Biol. Sci.*, **17**, 55–98

——(forthcoming). 'Some reflections on the history of particle physics in the 1950s', in Brown *et al.* (forthcoming)

Seabourne, M. and Lowe, R. (1977). *The English School, Vol. 2 (1870–1970)*, London

Secord, J.A. (1989). 'Extraordinary experiment: Electricity and the creation of life in Victorian England', in Gooding *et al.* (1989), pp. 337–83

Seidel, R.W. (1983). 'Accelerating science: The postwar transformation of the Lawrence Radiation Laboratory', *Hist. Stud. Phys. Sci.*, **13**, 375–400

——(1986). 'A home for big science: The Atomic Energy Commission's laboratory system', *Hist. Stud. Phys. Sci.*, **16**, 135–75

——(forthcoming). 'The postwar political economy of high energy physics', in Brown *et al.* (forthcoming)

Seriff, A.J., Leighton, R.B., Hsio, C., Cowan, E.W. and Anderson, C.D. (1950). 'Cloud-chamber observations of the new unstable cosmic-ray particles', *Phys. Rev.*, **78**, 290

Service, A. (Ed.) (1975). *Edwardian Architecture and its Origins*, London

Settle, T. (1961). 'An experiment in the history of science', *Science, N.Y.*, **133**, 19–23

——(1983). 'Galileo and early experimentation', in Aris *et al.* (1983), pp. 3–20

Sfez, L. (1981). *Critique de la décision*, Paris

Shapin, S. and Schaffer, S. (1985). *Leviathan and the Air-pump: Hobbes, Boyle and the Experimental Life*, Princeton

Sharp, E. (1926). *Hertha Ayrton, 1854–1923*, London

Shryock, R.H. (1948). 'American indifference to basic science during the nineteenth century', *Arch. Int. Hist. Sci.*, **2**, 50–65

Siegfried, R. (1959). 'The chemical philosophy of Humphry Davy', *Chymia*, **5**, 193–201

Simpson, A.D.C. (Ed.) (1982). *Joseph Black 1728–1799: A Commemorative Symposium*, Edinburgh

Six, J. and Artru, S. (1982). 'An essay chronology of particle physics until 1965', *J. Phys.*, Colloque C8, supplément au No. 12, **43**, C8-465–C8-496

Smeaton, W.A. (1965–66). 'The portable laboratories of Guyton de Morveau', *Ambix*, **13**, 84–91

Smith, S.A. (1976). 'Alfred Waterhouse: civic grandeur', in Fawcett (1976), pp. 102–21

Spieker, P. (1888). 'Sternwarten und andere Observatorien', in Durm *et al.* (1888), pp. 474–567

Stansfield, D.A. (1984). *Thomas Beddoes M.D. 1760–1808*, Dordrecht, Boston, Lancaster

Statham, W.E. [1862]. *A Catalogue of Amusing and Instructive Articles ... Manufactured and Sold by W.E. Statham*, London

——(1903). *Second Steps in Chemistry*, London

Staudenmaier, J.M. (1985). *Technology's Storyteller*, Cambridge

Stephen, J. [1887]. *Playground of Science*, London

Stock, J.E. (1811). *Memoirs of the life of Thomas Beddoes, M.D. with an Analytical Account of His Writings*, London

Stoops, R. (1961). *The Quantum Theory of Fields*, New York and London

Storr, R.J. (1953). *The Beginnings of Graduate Education in America*, Chicago

Streatfield, F.W. (1912). 'The City & Guilds of London Technical College, Finsbury', *Chem. World*, **1**, 373–7

Struve, F.G.W. (1825). *Beschreibung des grossen Refractors von Fraunhofer*, Dorpat

——(1845). *Description de l'observatoire astronomique central de Pulkowa*, 2 volumes, St. Petersburg

——(1847). *Etudes d'astronomie stellaire*, St. Petersburg

Swenson, L.S. (1972). *The Ethereal Aether: A History of the Michelson-Morley-Miller Aether Drift Experiments, 1880–1930*, Austin

Swinbank, P. (1982). 'Experimental science in the University of Glasgow at the time of Joseph Black', in Simpson (1982), pp. 23–35

Teague, J. (1980). *The City University. A History*, London

Thompson, S.P. [1901]. *Report to the Committee on the Status and Outlook of the City and Guilds Technical College, Finsbury*, London

Thomson, G.P. (1964). *J.J. Thomson and the Cavendish Laboratory in His Day*, Garden City

Thomson, J. (1832–41). *An Account of the Life, Lectures, and Writings of William Cullen, M.D.*, 2 volumes, Edinburgh and London

Thomson, J.J. (1936). *Recollections and Reflections*, London

Thomson, T. (1830). *The History of Chemistry*, 2 volumes, London

Thomson, W. and Tait, P.G. (1912). *Treatise on Natural Philosophy*, 2nd edn, London

Thorpe, T.E. (1896). *Humphry Davy, Poet and Philosopher*, London

Tobey, R. (1971). *The American Ideology of National Science*, Pittsburg

Todd, A.C. (1967). *Beyond the Blaze*, Truro

Trowbridge, J. (1871). 'A new form of galvanometer', *Am. J. Sci.*, 3rd series, **2**, 118–20

——(1872). 'Remarks on animal electricity', *Proc. Am. Acad. Arts Sci.*, **8**, 344–7

——(1877?). *The Endowment of a Physical Laboratory at Harvard College*, Cambridge, Mass.

——(1884). *The New Physics*, New York

——(1885). 'The Jefferson Physical Laboratory', *Science, N.Y.*, **5**, 229–31

——(1897). 'The energy conditions necessary to produce the Rontgen rays', *Proc. Am. Acad. Arts Sci.*, **32**, 255–65

Turner, G.L'E. (Ed.) (1976). *Patronage of Science in the Nineteenth Century*, Leyden

Tweney, R.D. (1985). 'Faraday's discovery of induction: A cognitive approach', in Gooding and James (1985), pp. 189–209

Vernon, K.D.C. (1963). 'The foundation and early years of the Royal Institution', *Proc. Roy. Inst.*, **39**, 364–402

Veyne, P. (1979). *Comment on écrit l'histoire*, Paris

Veysey, L.R. (1965). *The Emergence of the American University*, Chicago

Walker, E.G. (1933). 'Finsbury Technical College', *Central*, **30**, 35–48

——(1950). 'The last years of Finsbury Technical College's Old Students' Association', *Central*, 5–7

Walker, W. (1862). *Distinguished Men of Science*, London

Webster, D.L. (1938). 'Contributions of Edwin Herbert Hall to the teaching of physics', *Am. Phys. Teacher*, **6**, 15

Weinberg, A.R. (1963). 'Criteria for scientific choice', *Minerva*, **1**, 159–71

——(1967). *Reflections on Big Science*, Oxford

Weindling, P. (1983). 'The British Mineralogical Society: A case study in science and social development', in Inkster and Morrell (1983), pp. 120–50

West, F. (1833). *A Short Account of Intellectual Toys*, London

Wien, W. (1926). 'Das Physikalische Institut und das Physikalische Seminar', in Muller (1926), pp. 207–11

Wightman, A.S. (forthcoming). 'The general theory of quantized fields in the 1950s', in Brown *et al.* (forthcoming)

Williams, I. (1968). 'The Edgeworths and practical education', *School Sci. Rev.*, **7**, 21–33

Williams, L.P. (1983). 'What were Ampère's earliest discoveries in electrodynamics?', *ISIS*, **74**, 492–508

——(1985). 'Faraday and Ampère: A critical dialogue', in Gooding and James (1985), pp. 83–104

—— *et al.* (1971). *The Selected Correspondence of Michael Faraday*, 2 volumes, Cambridge

Williams, M.E.W. (1981). *Attempts to Measure Annual Stellar Parallax: Hooke to Bessel*, University of London (Imperial College), PhD thesis

Willson, R.W. (1890). 'The magnetic field in the Jefferson Physical Laboratory', *Am. J. Sci.*, 3rd series, **39**, 87–93, 456–70

Wilson, D. (1983). *Rutherford: Simple Genius*, London

Wohl, A.S. (1983). *Endangered Lives. Public Health in Victorian Britain*, London

Wolf, C.J.E. (1902). *Histoire de l'observatoire de Paris de sa fondation à 1793*, Paris

Woolf, H. (1959). *The Transits of Venus*, Princeton

Yuan, L.C.L. (Ed.) (1965). *Nature of Matter: Purposes of High Energy Physics*, Springfield

—— and Blewett, J.P. (1961). *Experimental Requirements for a 300 to 1000-BeV Accelerator*, Brookhaven

Zichichi, A. (Ed.) (1970). *Subnuclear Phenomena*, New York

Index*

---

* Laboratories, constituent colleges, etc., belonging to institutions are indexed under that institution.